International Conference on Quantum Optics and Quantum Information 2013

(icQoQi)

Journal of Physics: Conference Series Volume 553

Pahang, Malaysia
3 – 4 December 2013

ISBN: 978-1-63439-665-3
ISSN: 1742-6588

Printed from e-media with permission by:

Curran Associates, Inc.
57 Morehouse Lane
Red Hook, NY 12571

Some format issues inherent in the e-media version may also appear in this print version.

Printed by Curran Associates, Inc. (2015)

For permission requests, please contact the Institute of Physics
at the address below.

Institute of Physics
Dirac House, Temple Back
Bristol BS1 6BE UK

Phone: 44 1 17 929 7481
Fax: 44 1 17 920 0979

techtracking@iop.org

Additional copies of this publication are available from:

Curran Associates, Inc.
57 Morehouse Lane
Red Hook, NY 12571 USA
Phone: 845-758-0400
Fax: 845-758-2634
Email: curran@proceedings.com
Web: www.proceedings.com

TABLE OF CONTENTS

Preface

Quantum Information can be understood as being naturally derived from a new understanding of information theory when quantum systems become information carriers and quantum effects become non negligible. Experiments and the realization of various interesting phenomena in quantum information within the established field of quantum optics have been reported, which has provided a very convenient framework for the former. Together, quantum optics and quantum information are among the most exciting areas of interdisciplinary research in modern day science which cover a broad spectrum of topics, from the foundations of quantum mechanics and quantum information science to the introduction of new types of quantum technologies and metrology.

The International Conference on Quantum Optics and Quantum Information (icQoQi) 2013 was organized by the Faculty of Science, International Islamic University Malaysia with the objective of bringing together leading academic scientists, researchers and scholars in the domain of interest from around the world to share their experiences and research results about all aspects of quantum optics and quantum information.

While the event was organized on a somewhat modest scale, it was in fact a rather fruitful meeting for established researchers and students as well, especially for the local scene where the field is relatively new. We would therefore, like to thank the organizing committee, our advisors and all parties for having made this event successful and last but not least would extend our sincerest gratitude to IOP for publishing these selected papers from icQoQi2013 in Journal of Physics: Conference Series.

Peer review statement

All papers published in this volume of *Journal of Physics: Conference Series* have been peer reviewed through processes administered by the proceedings Editors. Reviews were conducted by expert referees to the professional and scientific standards expected of a proceedings journal published by IOP Publishing.

Exact solitonic solutions for optical media with $\chi^{(2)}$ nonlinearity and PT-symmetric potentials

F. Kh. Abdullaev, B. A. Umarov

Department of Physics, Kulliyyah of Sciences, International Islamic University of Malaysia, Jalan Istana, 25200, Bandar Indera Mahkota Kuantan, Malaysia

E-mail: fatkhulla@iium.edu.my

Abstract. We study the $\chi^{(2)}$ system of equations with the PT-symmetric potentials. We obtain exact solitary wave solutions for this system for some forms of the PT-symmetric potentials.

1. Introduction

Recently a great attention has been devoted to the investigation of quantum and classical systems with PT-symmetric non-Hermitian Hamiltonians. It was shown in the seminal paper [1] that such type of Hamiltonians can have real eigenvalues. This result has growing applications in optics, Bose-Einstein condensates, plasmonics etc. In particular in optics due to the analogy between the Schrodinger equation and the paraxial wave equation for optical beams the PT-symmetry imposes the condition on the complex refractive index $n(x) = n_r(x) + in_i(x)$: even in space of a real part of refractive index $n_r(x) = n_r(-x)$ and odd in space imaginary part $n_i(x) = -n(-x)$. Recently effects of PT- symmetry have been observed in optical experiments[2]. The nonlinearity leads to new effects in the PT- symmetric systems, such as the nonreciprocity[3, 4], smoothing the spectral singularity in transmission [5] etc. It was shown recently that in nonlinear wave equations with PT- symmetric extension solitonic solutions can exist [6, 7, 8, 9, 10]. Particular interest represents search of exact solutions for the nonlinear waves in media with PT-symmetric modulations of parameters. In the case of the media with the Kerr nonlinearity described by the nonlinear Schrödinger equation with inhomogeneous in space loss/gain parameters such solutions were found for linear PT-potentials in the works [6, 7, 11, 12, 13, 14], for nonlinear PT-potentials - in the work[7, 8]. Recent numerical simulations of the $\chi^{(2)}$ system with PT - symmetric potential have showed the existence of stable bright solitons[15]. In this work we will study the $\chi^{(2)}$ system with PT - symmetric potentials, describing wave processes in quadratically nonlinear media and we will find the exact solitary solutions describing localized wave packets in such systems.

2. The model

Let us consider the $\chi^{(2)}$ system describing the FH and SH propagation in quadratically nonlinear media with PT-symmetric potentials[16]

$$iu_z + d_1 u_{xx} + V_1(x)u + iW_1(x)u = u^* v, \quad (1)$$

$$iv_z + d_2 v_{xx} + qv + V_2(x)v + iW_2(x)v = u^2, \quad (2)$$

with $d_{1,2}$ are diffraction coefficients, q is the phase missmatch. Here $V_{1,2}(x)$ are real parts of the refraction index. PT-symmetry impose restriction on $V_{1,2}$ to be even functions of x and and on the imaginary potential $W_{1,2}(x)$ describing the inhomogeneous in space gain/loss, to be odd functions of x. Introducing representation

$$u(x,z) = U(x)e^{-i\omega z}, v(x,z) = V(x)e^{-2i\omega z},$$

we obtain the system of equations for $U(x)$ and $W(x)$

$$\omega U + d_1 U_{xx} + V_1(x)U + iW_1(x)U = U^* V, \quad (3)$$
$$\sigma V + d_2 V_{xx} + V_2(x)V + iW_2(x)V = U^2, \quad (4)$$

where $\sigma = 2\omega + q$. It is useful to introduce the amplitude-phase variables by

$$U(x) = \rho_1(x)e^{i\theta(x)},\ V(x) = \rho_2(x)e^{2i\theta(x)}.$$

Then we obtain the system:

$$\omega\rho_1 + d_1\rho_{1,xx} - d_1\rho_1(\theta_x)^2 + V_1(x)\rho_1 = \rho_1\rho_2, \quad (5)$$
$$\sigma\rho_2 + d_2\rho_{2,xx} - 4d_2\rho_2(\theta_x)^2 + V_2(x)\rho_2 = \rho_1^2, \quad (6)$$
$$W_1(x)\rho_1^2 = -d_1(\rho_1^2\theta_x)_x, \quad (7)$$
$$W_2(x)\rho_2^2 = -2d_2(\rho_2^2\theta_x)_x. \quad (8)$$

To find the exact solutions of the system (5) we will apply the inverse engineering approach i.e. construct the potentials admitting the exact solutions[17, 7].

3. Bright soliton solutions

First we consider the case of localized solutions in the form of bright-bright solitonic solutions.

1. Let look for solutions of the form:

$$\rho_1 = A\mathrm{sech}^2(rx),\ \rho_2 = B\mathrm{sech}^2(rx). \quad (9)$$

This form, for constant parameters case, has been introduced by Karamzin-Sukhorukov[18]. For the phase we impose the form:

$$\theta_x(x) = C\mathrm{sech}(rx). \quad (10)$$

Then we obtain from (5) for the $W_1(x)$

$$W_1(x) = 5Cd_1 r\mathrm{sech}(rx)\tanh(rx) = W_{01}\mathrm{sech}(rx)\tanh(rx). \quad (11)$$

Analogously we have found

$$W_2(x) = 10Cd_2 r\mathrm{sech}(rx)\tanh(rx) = W_{02}\mathrm{sech}(rx)\tanh(rx).$$

Then

$$C = \frac{W_{01}}{5d_1 r} = \frac{W_{02}}{10d_2 r}.$$

We obtain the relation between W_{01}, W_{02}

$$W_{02} = 2\frac{d_2}{d_1}W_{01}.$$

From equations for amplitudes we have

$$V_i = V_{0i}\text{sech}^2(rx),\ \omega = -4d_1r^2, \sigma = -4d_2r^2, q = 4r^2(2d_1 - d_2). \tag{12}$$

$$B = (V_{01} - 6d_1r^2 - \frac{W_{01}^2}{25r^2d_1}),\ A^2 = B(V_{02} - 6d_2r^2 - \frac{4d_2W_{01}^2}{25r^2d_1^2}). \tag{13}$$

The phase is:

$$\theta(x) = \frac{2W_{01}}{5r^2d_1}\text{arctg}(\tanh(rx/2)). \tag{14}$$

For example if $r = 1, d_1 = 1/2, d_2 = 1/4$ we obtain that the phase mismatch should be $q = 3$. Taking into account the condition $A^2 > 0$, dynamical breaking of the PT- symmetry occurs, when $B > 0$ and the strength of imaginary part of potential excess the value

$$W_{1c}(x) = 5r\sqrt{d_1(V_{01} - 6d_1r^2)}, \tag{15}$$

or when $B < 0$ and

$$W_{1c}(x) < 5r\sqrt{d_1(V_{01} - 6d_1r^2)}, \tag{16}$$

with $V_{01} = 2V_{02}, d_1 = 2d_2$. Existence and stability of the solution when the conditions above are not fulfilled requires a separate consideration.

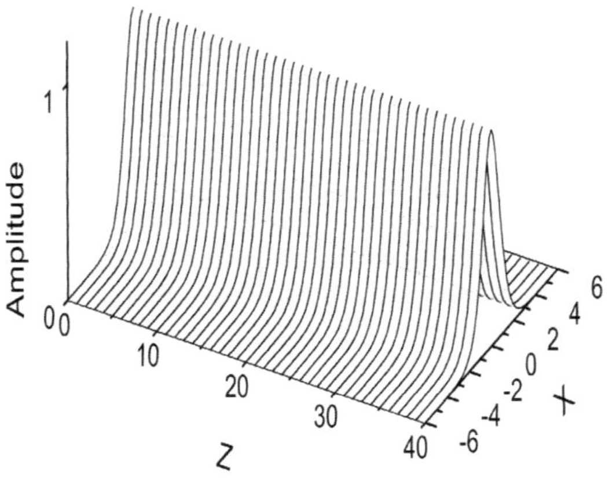

Figure 1. Evolution of the amplitude $|u(x,z)|$ with initial conditions in form of exact solution (9-10). Parameters: $A = 1$, $B = 1$, $r = 1$, $d_1 = 0.5$, $d_2 = 0.25$, $W_{01} = 1$.

2. Second choice of the solution is:

$$\rho_1(x) = A\text{sech}(rx),\ \rho_2 = B\text{sech}^2(rx). \tag{17}$$

This form for constant parameters of $\chi^{(2)}$ system has been introduced in [19]. Again we will assume the functional dependence for the phase gradients in the form (10). Then we obtain for $W_i(x)$ relations:

$$\begin{aligned} W_1 &= 3Cd_1r\text{sech}(rx)\tanh(rx) = W_{01}\text{sech}(rx)\tanh(rx), \\ W_2 &= 10Cd_2r\text{sech}(rx)\tanh(rx) = W_{02}\text{sech}(rx)\tanh(rx), \\ C &= \frac{W_{01}}{3rd_1} = \frac{W_{02}}{10d_2r}. \end{aligned} \tag{18}$$

Then we have

$$W_{02} = \frac{10 d_2 W_{01}}{3 d_1}.$$

It can be shown, that

$$\omega = -d_1 r^2.$$

For soliton amplitudes we find respectively:

$$B = V_{01} - 2d_1 r^2 - \frac{W_{01}^2}{9r^2 d_1}, A^2 = B(\sigma + 4d_2 r^2). \tag{19}$$

For real potentials we have expressions

$$V_1(x) = V_{01}\mathrm{sech}^2(rx), V_2(x) = V_{02}\mathrm{sech}^2(rx,), \tag{20}$$

where

$$V_{02} = 6d_2 r^2 (1 + \frac{2W_{01}^2}{27 d_1^2 r^4}). \tag{21}$$

Again we have the dynamical breaking of the PT symmetry above threshold value for the strength of the imaginary part of potential W

$$W_{1c} = 3r\sqrt{d_1(V_{01} - 2d_1 r^2)}.$$

It is useful to consider the large negative mismatch case when $|q| >> 1$. Then $v \approx -u^2/|q|$ and the system is reduced to the PT-symmetric extension of the single NLS equation(the so-called a cascading limit case) The solitonic solution for the potentials $V = V_0\mathrm{sech}^2(x), W = W_0\mathrm{sech}(x)\tanh(x)$ is well known[6] and coincides with the solution (19), when $|q| >> 1, r = 1, d_1 = 1$.

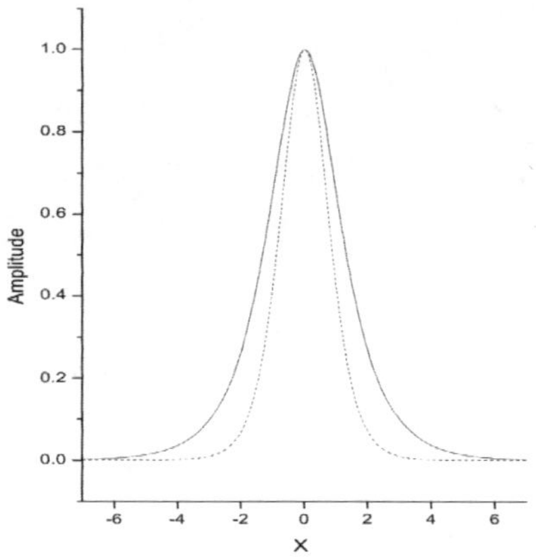

Figure 2. Amplitude of the exact solution (16) $|u(x)|$ solid line and $|v(x)|$ dashed line at $z = 0$. Parameters: $A = 1$, $B = 1$, $r = 1$, $d_1 = 0.5$, $d_2 = 0.25$, $W_{01} = 1$.

3.Third choice of solutions is:

$$\rho_1(x) = A\mathrm{sech}(rx)\tanh(rx)e^{i\theta(x)+i\pi/2}, \ \rho_2(x) = B\mathrm{sech}^2(rx)e^{2i\theta(x)}. \tag{22}$$

This type of solitonic solutions for an homogeneous $\chi^{(2)}$ media is considered in [20, 21]. Let take the phase of the form

$$\theta_x = C\mathrm{sech}(rx)\tanh(rx).$$

Looking for the imaginary part of potential of the form

$$W_1(x) = W_1(\mathrm{sech}(rx) - 2\mathrm{sech}^3(rx)), W_2(x) = W_2\mathrm{sech(rx)}(5 - 6\mathrm{sech}^2(rx))$$

we obtain

$$C = \frac{W_1}{3d_1 r} = \frac{W_2}{2d_2 r} = \sqrt{\frac{|V_1^{(2)}|}{d_1}} = \sqrt{\frac{|V_2^{(2)}|}{4d_2}}.$$

Also we find

$$\begin{aligned} A^2 &= -(\sigma + 4d_2 r^2)B, B = (-(V_1^{(1)} + V_1^{(2)}) + 6d_1 r^2), \\ V_i(x) &= V_i^{(1)}\mathrm{sech}^2(x) + V_i^{(2)}\mathrm{sech}^4(x), i = 1, 2; \theta(x) = -\frac{W_1}{3d_1 r^2}\mathrm{sech}(rx), \ \omega = -d_1 r^2, \\ \sigma &= 2d_2 r^2 - (V_2^{(1)} + V_2^{(2)}). \end{aligned} \tag{23}$$

So, as one can note, this is the example of exact solution in the complex non PT symmetric potential, since both real and imaginary parts of the potential are even functions of the x variable.

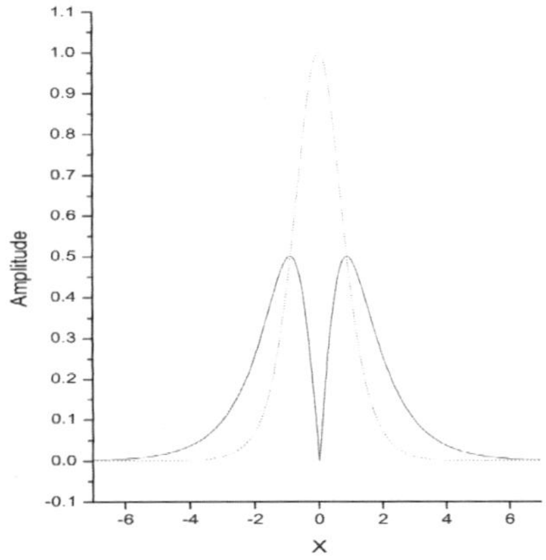

Figure 3. Amplitude of the exact solution (21) $|u(x)|$ solid line and $|v(x)|$ dashed line at $z = 0$. Parameters: $A = 1$, $B = 1$, $r = 1$, $d_1 = 0.5$, $d_2 = 0.25$, $W_{01} = 1$.

4. Conclusion

. We have found exact solitary solutions for quadratically nonlinear media with PT- symmetric potentials. The solutions were in the form of bright-bright solitons. Also the example of exact solution in non PT symmetric potential is given. The investigation of stability of obtained exact solutions is important problem and requires a separate analysis.

Acknowledgments

Authors acknowledge partial financial support from FRGS grant 13-092-0333 MOHE Malaysia.

References

[1] Bender C M and Boettcher S 1998 *Phys. Rev. Lett.* **80**, 5243
[2] Guo A, Salamo G J, Duchesne D, Morandotti R, Volatier- Ravat M, Aimez V, Siviloglou S A and Christodoulides D N 2009 *Phys. Rev. Lett.* **103**, 093902

[3] Ramezani H,Kottos T, El-Ganainy R and Christodoulides D N 2008 *Phys. Rev.* A **82** 043803
[4] Sukhorukov A A, Zhiyong Xu and Kivshar Yu 2010 *Phys. Rev* A **82** 013833
[5] Liu X, Gupta S D and Agarwal 2013 *arXiv* 1310.8282
[6] Musslimani Z H, Markis K S, El-Ganainy R and Christodoulides D N 2008 *Phys. Rev. Lett.* **100**, 030402 ; *J. Phys.* A **41**, 244019
[7] Abdullaev F Kh, Konotop V V, Salerno M and Yulin A V 2010 *Phys. Rev.* E **82**, 056606
[8] Abdullaev F Kh, Kartashov Y Y, Konotop V V and Zezyulin D A 2011 *Phys. Rev.* A **83**, 041805
[9] Zezyulin D A, Kartashov Y V and Konotop V V, 2011 *Eur. Phys. Lett.* **96**, 64003
[10] Driben R and Malomed B A 2011 *Opt.Lett.* **36**, 4323
[11] Salerno M arXiv:1306.3643v1
[12] Xu H, Kevrekidis P G, Zhou Q, Frantzeskakis D J, Achilleos V and Carretero-Gonzalez R arXive 1310.7652v1
[13] Mayteevarunyoo T, Malomed B. A. and Reoksabutr A 2013 *Phys.Rev.* E **88**, 022919
[14] Hanif W, Karjanto N, Malomed B. A. and Susanto H 2014 *arXiv* 4241v1
[15] Moreira F, Abdullaev F Kh, Konotop V V and Yulin A, 2012 *Phys.Rev.* A **86** 053815
[16] Parker D F and Tsoy E N 1999 *J.Eng.Math.* **36**, 149
[17] Brazhnyi V A and Konotop V V 2004 *Mod. Phys. Lett.* B **18**, 627
[18] Karamzin Y N and Sukhorukov A P, 1974 *JETP Lett.* **20** 339
[19] Werner M J and Drummond P D 1994 *Opt.Lett.* **19**, 613
[20] Werner M J and Drummond P D 1994 *JOSA* B **10** 2390
[21] Menyuk C R, Schiek R and Torner L 1994 *JOSA* B **11**, 2434

Journal of Physics: Conference Series **553** (2014) 012002 doi:10.1088/1742-6596/553/1/012002

Pure tripartite entanglement types based on spectra of reduced density matrices

Saeid Molladavoudi*[1], Hishamuddin Zainuddin[1,2] and Chan Kar Tim[2]

[1] Laboratory of Computational Sciences and Mathematical Physics, Institute for Mathematical Research, Universiti Putra Malaysia, 43400 UPM Serdang, Selangor, Malaysia
[2] Department of Physics, Faculty of Science, Universiti Putra Malaysia, 43400 UPM Serdang, Selangor, Malaysia

E-mail: *saeid.molladavoudi@gmail.com, hisham@upm.edu.my, chankt@upm.edu.my

Abstract. Given a specific ordered spectra of (shifted) one-qubit reduced density matrices, we discuss that the pure tripartite entanglement for a system consisting of three qubits is determined by only two local unitary invariant polynomials.

1. Introduction

Generally, for a composite quantum system in its pure state the classification of entanglement classes amounts to partitioning the associated projective space into orbits of the Local Unitary (LU) operations. However, according to a general principle in mathematics in order to simplify the problem one has to exploit the existing symmetries whenever possible. In our particular case of interest, the LU group is a Lie group K, which acts on the pure states space of a composite quantum system. This group action is a *proper* action, since the LU group K is a compact Lie group. In addition, the multi-particle pure states space $\mathbb{P}(\mathcal{H})$ is a Kähler manifold, for which the Kähler structure is induced from the Hermitian inner product on the Hilbert space $\mathcal{H}$.

The symmetry is then the conservation of the spectra of each qubit's reduced density matrix under the local unitary operation at each site. The appropriate mathematical machinery to exploit this symmetry is encoded in the Symplectic Geometry, in which there exists a map, so called the *momentum map*, whose components are preserved under the action of compact Lie group. In [1, 2], it is shown that the action of the LU group K on the Kähler manifold $\mathbb{P}(\mathcal{H})$ is symplectic, i.e. the symplectic structure ω (a closed non-degenerate 2-form) is preserved under the action of K. Also, the associated momentum map is given by sending each pure multipartite state to a collection of (shifted) one-body reduced density matrices.

More precisely, if a compact Lie group G acts on a symplectic manifold (N, Ω) properly by preserving its symplectic structure, then there exists the momentum map $\mu : N \to \mathfrak{g}^*$, where $\mathfrak{g}^*$ is the dual space for the Lie algebra $\mathfrak{g}$ of the Lie group G. There exists an isomorphism $\mathfrak{g} \cong \mathfrak{g}^*$ by defining a G-invariant inner product on the Lie algebra. Then the quadruple (N, Ω, G, μ) is called a *Hamiltonian G-space*, which can be reduced to a generally topological quotient space $N_\alpha = \mu^{-1}(\alpha)/K_\alpha$, where $\alpha \in \mathfrak{g}^*$, by exploiting the existing symmetry encoded in the components of the momentum map μ. Here the subgroup K_α is the stabilizer of α by the coadjoint action of the Lie group G on $\mathfrak{g}^*$.

In the current paper, we review this reduction procedure for the Hamiltonian K-space $(\mathbb{P}(\mathcal{H}), \omega, K, \mu)$ for the case of pure tripartite states under the LU group K-action. To do so we employ the image under the Hilbert map of the algebra of local unitary invariant polynomials, which is finitely generated [4], and the generalized Schmidt decomposition for pure tripartite states for three qubits [5], whose components are in one-to-one correspondence with local unitary invariant polynomials. For more details and the original results the readers can refer to [3].

2. Local Unitary Operations

For a composite quantum system consisting of N (distinguishable) particles the LU group equals to $K = SU(n_1) \times SU(n_2) \times \cdots SU(n_N)$, where n_j denotes the $\dim(\mathcal{H}_j)$ with $\mathcal{H}_j = \mathbb{C}^{n_j}$ for the jth particle. The multipartite pure states space in then the projective Hilbert space $\mathbb{P}(\mathcal{H})$, as a Kähler manifold, where $\mathcal{H} = \otimes_{j=1}^{N} \mathcal{H}_j$. The LU group K acts on the Hilbert space $\mathcal{H}$ as follows

$$k \,.\, \phi = k_1\phi_1 \otimes k_2\phi_2 \otimes \cdots \otimes k_N\phi_N, \tag{1}$$

where $k = (k_1, k_2, \cdots, k_n) \in K$ and $\phi = \phi_1 \otimes \phi_2 \otimes \cdots \otimes \phi_N \in \mathcal{H}$, with normalized $\phi_j \in \mathcal{H}_j$. The action of the LU group K on the manifold $\mathbb{P}(\mathcal{H})$ is induced from its action (1) on $\mathcal{H}$, which is given by conjugation, i.e.

$$K \times \mathbb{P}(\mathcal{H}) \to \mathbb{P}(\mathcal{H}), (k, \rho) \mapsto k \,.\, \rho = k\,\rho\,k^{-1}, \tag{2}$$

for all $k \in K$ and $\rho \in \mathbb{P}(\mathcal{H})$. Recall that the projection $\pi : \mathcal{H} \to \mathbb{P}(\mathcal{H})$ is given by sending each state $\phi \in \mathcal{H}$ to the corresponding trace-class, rank-one Hermitian operator $\rho = \phi\,\phi^*$, where ϕ^* is the complex conjugate of the multipartite pure state ϕ.

The orbits of the K-action on the manifold $\mathbb{P}(\mathcal{H})$ are given by the sets $K \,.\, \rho = \{k \,.\, \rho : \forall k \in K\}$. This establishes an equivalence relation in $\mathbb{P}(\mathcal{H})$, i.e. two states ρ_1 and ρ_2 are equivalent if and only if there exists an element $k \in K$, such that $\rho_2 = k \,.\, \rho_1$, i.e. ρ_1 and ρ_2 belong to the same orbit. According to the Hilbert's theorem [7] for each orbit $K \,.\, p$, there exists an LU-invariant polynomial which is constant on it and the number of linearly independent LU-invariant polynomials is finite. Since the LU group is a compact Lie group the space of orbits, which is denoted by $\mathbb{P}(\mathcal{H})/K$, is a Hausdorff space. Then the topological space of orbits $\mathbb{P}(\mathcal{H})/K$ is mapped into a subset of the Euclidean space $\mathbb{R}^m$, where m is the number of basis for the algebra of LU-invariant polynomials.

For the case of pure tripartite states, i.e. $N = 3$, the LU-invariant polynomials are related to the generalized Schmidt decomposition [5]. In fact, every pure three-qubit state $\phi \in \mathcal{H}$ can be transformed to the following normal form

$$|\phi_0\rangle = \lambda_0|000\rangle + \lambda_1\, e^{i\phi}|100\rangle + \lambda_2|101\rangle + \lambda_3|110\rangle + \lambda_4|111\rangle, \tag{3}$$

by local unitary operations. The normal form (3), with $\lambda_k \geq 0$, for $k = 0, \cdots, 4$, and $\sum_{k=0}^{4} \lambda_k^2 = 1$ with $\phi \in [0, \pi]$ is non-unique up to the permutation of qubits. In [5, 6], the LU-invariant polynomials are also constructed in terms of LU-invariant coefficients in the normal form (3), i.e.

$$\begin{aligned} I_1 &= |\lambda_1\lambda_4 e^{i\phi} - \lambda_2\lambda_3|^2, \quad I_2 = \lambda_0^2\lambda_2^2, \quad I_3 = \lambda_0^2\lambda_3^2, \quad I_4 = \lambda_0^2\lambda_4^2, \\ I_5 &= \lambda_0^2(I_1 + \lambda_2^2\lambda_3^2 - \lambda_1^2\lambda_4^2), \end{aligned} \tag{4}$$

which satisfy some equalities and inequalities. These constraints determine image of the orbit space $\mathbb{P}(\mathcal{H})/K$ as a semi-algebraic subset of the Euclidean space $\mathbb{R}^m$, where $m = 5$, in terms of LU-invariant polynomials [8].

2.1. *Momentum Map*

Recall that the pure multipartite state space $\mathbb{P}(\mathcal{H})$ is a Kähler manifold, for which the symplectic structure is given by the imaginary part of the Hermitian inner product on $\mathcal{H}$, or

$$\omega_\rho(\hat{A},\hat{B}) := \frac{\mathrm{i}}{2}\mathrm{Tr}(\rho\,[A,B]), \tag{5}$$

for two Hermitian operators $A, B \in \mathfrak{su}^*(\mathcal{H})$. In Eq. (5) $\hat{A},\hat{B} \in T_\rho(\mathbb{P}(\mathcal{H}))$ are the fundamental vector fields of the action of the Lie algebra $\mathfrak{su}(\mathcal{H})$ on the manifold given and $[\,.\,,\,.\,]$ denotes the commutator between the operators. As mentioned above, the LU-group action on the complex projective manifold $\mathbb{P}(\mathcal{H})$ is symplectic, and since the LU group K is compact, there exists a momentum map $\mu : \mathbb{P}(\mathcal{H}) \to \mathfrak{k}^*$, which is defined as follows [1, 2]

$$\mu(\rho) = \bigoplus_{l=1}^{N}(\rho^{(l)} - \frac{1}{n_l}\mathbb{1}_{n_l}) \in \mathfrak{k}^*, \tag{6}$$

where $\rho^{(l)}$, for $l = 1,\cdots,N$, is the lth-qubit reduced density matrix.

3. Symmetry reduction of a Hamiltonian system

Let us first briefly review the symmetry reduction of a Hamiltonian G-manifold (N,Ω,G,μ), when a compact Lie group G acts symplectically on a symplectic manifold (N,Ω). Recall that the image of the manifold N under the *invariant momentum map* $\psi : N \to \mathfrak{g}^*/G, \rho \mapsto \psi(\rho) \equiv \alpha_0$ is a convex polytope Δ which is a subset of the positive Weyl chamber $\mathfrak{g}^*/G$ [9]. The relation between $\mu(\rho) \equiv \alpha \in \mathfrak{g}^*$ and $\psi(\rho) \equiv \alpha_0 \in \mathfrak{g}^*/G$ is that α_0 the intersection of the coadjoint orbit $\mathcal{O}_\alpha = \{\mathrm{Ad}_g\alpha : \forall g \in G\}$, where $\mathrm{Ad}_g\alpha = g^{-1}\,\alpha\, g$, with $\mathfrak{g}^*/G$.

In general, the resulting reduced space, which is denoted by $N_{\mathcal{O}_\alpha} = \psi^{-1}(\alpha_0)/K$, is topologically a stratified space such that each piece (stratum) is a symplectic manifold and it contains all the points in the fiber $\psi^{-1}(\alpha_0)$, for $\alpha_0 \in \Delta$, whose stabilizers are conjugate with a subgroup of the Lie group G. For more details the interested readers can refer to [10, 11]. The crucial in our analysis is the properness of the Lie group and equivariance property of the momentum map. The following theorem will be used in continue to obtain the image of the reduced space of Hamiltonian G-manifold in terms of the Hilbert map.

Theorem 3.1 *[10] Let (N,Ω,G,μ) be a Hamiltonain G-manifold. The space of orbits M/G is embedded into a Euclidean space $\mathbb{R}^m$ by the Hilbert map $\sigma : N/G \to \mathbb{R}^m$, where m is the number of bases of the G-invariant polynomials. The embedding of reduced space $N_{\mathcal{O}_{\alpha_0}} = \psi^{-1}(\mathcal{O}_{\alpha_0})/G$, for $\alpha_0 \in \mathfrak{g}^*/G$, is obtained by restriction of the map σ to $N_{\mathcal{O}_{\alpha_0}}$ as a subspace of the Euclidean space $\mathbb{R}^m$.*

4. Three-qubit case

Recall that the three-qubit polytope Δ is determined by the following polygonal inequalities [12, 13]

$$\gamma_k \leq \sum_{j\neq k}\gamma_j + \frac{1}{2}, \tag{7}$$

where $-\frac{1}{2} \leq \gamma_l \leq 0$ for $l = 1,2,3$ are the minimal eigenvalues of matrices $\rho^{(k)} - \frac{1}{2}\mathbb{1}_2$. The momentum value of the normal form $\mu(\rho_0)$, where with $\rho_0 = \phi_0\phi_0^*$, amounts to finding these matrices by considering the Eq. (6). Furthermore, the value of invariant momentum map for the normal form (3), i.e. $\psi(\rho_0)$, can be obtained by diagonalization and re-ordering of the eigenvalues of matrices $\rho^{(k)} - \frac{1}{2}\mathbb{1}_2$. The resulting minimal eigenvalues are obtained in terms of the LU-invariant polynomials (4) as follows

$$2\gamma_k = -\sqrt{1-4c_k}, \quad c_k = \sum_{j=1,\, j\neq k}^{4} I_j. \tag{8}$$

By restriction of the Hilbert map image of the space of orbits $\mathbb{P}(\mathcal{H})/K$ to the reduced space $\psi^{-1}(\alpha_0)/K$, for $\alpha_0 \in \Delta$, its Hilbert map image under the LU-invariant polynomials is determined. Let us choose α_0 in the relative interior of the polytope Δ (7), with equal γ_ks (for simplicity) such that $-\frac{1}{6} < \gamma < 0$ (see [3] for further details). The restriction of LU-invariant polynomials (4) to the constraints obtained from Eqs. (8) lead to the following inequality

$$I_5^2 \leq \frac{1}{2}\left(1 - 4\left(I_4 + \gamma^2\right)\right)^3, \tag{9}$$

for a given $\alpha_0 = \psi(\rho_0) \in$ rel. int.$(\Delta) \cap (-\frac{1}{6} < \gamma < 0)$. In other words, the LU-invariant polynomials (I_4, I_5) such that they satisfy the inequality (9) are considered as the Hilbert map image of the reduced space of the Hamiltonian K-manifold $(\mathbb{P}(\mathcal{H}), \omega, K, \psi)$ for pure tripartite states.

5. Summary

In this paper we reviewed that by fixing the spectra of one-qubit (shifted) reduced density matrices, the entanglement types of pure tripartite states are determined in terms of two LU-invariant polynomials.

References

[1] Sawicki A, Huckleberry A and Kuś M 2011 *Communications in Mathematical Physics* **305**(2) 441–468 ISSN 0010-3616
[2] Sawicki A and Kuś M 2011 *Journal of Physics A: Mathematical and Theoretical* **44** 495301
[3] Molladavoudi S and Zainuddin H *Pure three-qubit state revisited: symplectic quotient under local unitary transformations*, submitted 2013
[4] Linden N and Popescu S 1998 *Fortschritte der Physik* **46** 567–578 ISSN 1521-3978
[5] Acín A, Andrianov A, Costa L, Jané E, Latorre J I and Tarrach R 2000 *Phys. Rev. Lett.* **85**(7) 1560–1563
[6] Acín A, Andrianov A, Jané E and Tarrach R 2001 *Journal of Physics A: Mathematical and General* **34** 6725
[7] Weyl H 1997 *The Classical Groups: Their Invariants and Representations* (Princeton, NJ: Princeton University Press)
[8] Walck S N, Glasbrenner J K, Lochman M H and Hilbert S A 2005 *Phys. Rev. A* **72**(5) 052324
[9] Kirwan F 1984 *Inventiones Mathematicae* **77**(3) 547–552 ISSN 0020-9910
[10] Sjamaar R and Lerman E 1991 *The Annals of Mathematics* **134** 375–422
[11] Ortega J P and Ratiu T S 2004 *Momentum Maps and Hamiltonian Reduction* (*Progress in Mathematics* vol 222) (Boston: Birkhaüser)
[12] Higuchi A, Sudbery A and Szulc J 2003 *Phys. Rev. Lett.* **90**(10) 107902
[13] Sawicki A, Walter M and Kuś M 2013 *Journal of Physics A: Mathematical and Theoretical* **46** 055304

Journal of Physics: Conference Series **553** (2014) 012003 doi:10.1088/1742-6596/553/1/012003

Quadratic Plus Linear Operators which Preserve Pure States of Quantum Systems: Small Dimensions

Mansoor Saburov

Department of Computational & Theoretical Sciences, Faculty of Science, International Islamic University Malaysia, P.O. Box, 141, 25710, Kuantan, Pahang, MALAYSIA

E-mail: msaburov@iium.edu.my, msaburov@gmail.com

Abstract. A mathematical formalism of quantum mechanics says that a pure state of a quantum system corresponds to a vector of norm 1 and an observable is a self-adjoint operator on the space of states. It is of interest to describe all linear or nonlinear operators which preserve the pure states of the system. In the linear case, it is nothing more than isometries of Hilbert spaces. In the nonlinear case, this problem was open. In this paper, in the small dimensional spaces, we shall describe all quadratic plus linear operators which preserve pure states of the quantum system.

1. Introduction

A physical system is generally described by three basic ingredients: states; observables; and dynamics (or law of time evolution) or, more generally, a group of physical symmetries. A classical description can be given in a fairly direct way by a phase space model of mechanics: states are points in a symplectic phase space, observables are real-valued functions on it, time evolution is given by a one-parameter group of symplectic transformations of the phase space, and physical symmetries are realized by symplectic transformations. A quantum description consists of a Hilbert space of states, observables are self adjoint operators on the space of states, time evolution is given by a one-parameter group of unitary transformations on the Hilbert space of states, and physical symmetries are realized by unitary transformations (for more details see [7]). Thus, in the mathematical formulation of quantum mechanics, the pure states correspond to vectors of norm 1. Consequently, the set of all pure states corresponds to the unit sphere in the Hilbert space.

The theory of linear preserver problems (LPPs) deals with characterizing linear maps that preserve certain properties of the spaces they act on (see [3]-[6] for a survey). For example, in the theory of quantum information, some of the most famous LPPs ask what a map must look like if it preserves pure states of the system. The classic result of LPPs says that linear operators which preserve pure states of an $n-$level quantum system are nothing else but unitary transformations on the $n-$dimensional Hilbert space.

In quantum mechanics, we know that a wavefunction of a multi-fermionic system is completely antisymmetric. Namely, this requirement is a way of more completely and rigorously stating the Pauli exclusion principle, the wavefunction for a system of two or more electrons must change sign any time we permute the coordinates of any two electrons. This principle is fulfilled in a multi-fermionic system. The best way to describe the wavefunction in the multi-fermionic system is a Slater determinant. We know that a determinant mapping between the matrix spaces is nonlinear and completely positive (for a detail see [1]). Therefore, the Slater determinant of the multi-fermionic system can be considered a nonlinear completely positive mapping. That's why it is of independent interest to study nonlinear preserver problems in quantum mechanics.

Nonlinear preserver problems (NPPs) were studied in the paper [8]. A notion of $q-$pure channels was introduced by means of quasi quantum quadratic operators which preserve (the Bloch sphere of) pure states of two-level quantum system. Moreover, it was given a description of all quadratic plus linear operators which leave the two-dimensional sphere invariant. In this paper, we are going to generalize this result (for more details see [8]).

2. Nonlinear Preserver Problems

We will use $\mathcal{H}_m = (\mathbb{R}^m, (\cdot,\cdot))$ to denote a real Hilbert space of (finite) dimension m with a standard inner product of two vectors $(x, y) = \sum\limits_{i=1}^{m} x_i y_i$. Throughout this paper, we assume that elements of $\mathbb{R}^m$ are column vectors $x = (x_1, \dots, x_m)^T \in \mathbb{R}^m$. Let $e_i = (\delta_{i1}, \delta_{i2}, \cdots, \delta_{im})^T$, $i = \overline{1,m}$ be the standard basis of $\mathbb{R}^m$ where δ_{ij} is Kronecker's delta symbol. We define a norm $\|x\| = \sqrt{(x,x)}$ associated with the inner product. Let $\mathcal{B}^m = \{x \in \mathbb{R}^m : \|x\| \leq 1\}$ be the unit ball and $\mathcal{S}^{m-1} = \{x \in \mathbb{R}^m : \|x\| = 1\}$ be the set (the unit sphere) of states of the quantum system.

Let $\mathbf{L} : \mathbb{R}^m \to \mathbb{R}^n$ be a linear operator and $L = (l_{ij})_{i,j=1}^{n,m}$ be its matrix with respect to the standard bases of the spaces $\mathbb{R}^m$ and $\mathbb{R}^n$. Then one has that

$$\mathbf{L}(x) = \left(\sum_{j=1}^{m} l_{1j} x_j, \sum_{j=1}^{m} l_{2j} x_j, \cdots, \sum_{j=1}^{m} l_{nj} x_j \right)^T. \tag{2.1}$$

If we define vectors $\mathsf{L}_j = (l_{1j}, l_{2j}, \cdots, l_{nj})^T \in \mathbb{R}^n$, $j = \overline{1,m}$ then the linear operator (2.1) takes the following form

$$\mathbf{L}(x) = \sum_{j=1}^{m} \mathsf{L}_j x_j. \tag{2.2}$$

Let $\mathbf{B} : \mathbb{R}^m \times \mathbb{R}^m \to \mathbb{R}^n$ be a symmetric bilinear operator. A quadratic operator $\mathbf{Q} : \mathbb{R}^m \to \mathbb{R}^n$ is defined as follows

$$\mathbf{Q}(x) = \mathbf{B}(x, x), \quad \forall\, x \in \mathbb{R}^m.$$

It is well-known that every quadratic operator $\mathbf{Q} : \mathbb{R}^m \to \mathbb{R}^n$ uniquely defines the symmetric bilinear operator $\mathbf{B} : \mathbb{R}^m \times \mathbb{R}^m \to \mathbb{R}^n$ associated with the given quadratic operator $\mathbf{Q} : \mathbb{R}^m \to \mathbb{R}^n$

$$\mathbf{B}(x, y) = \frac{1}{4}[\mathbf{Q}(x + y) - \mathbf{Q}(x - y)] = \frac{1}{2}[\mathbf{Q}(x + y) - \mathbf{Q}(x) - \mathbf{Q}(y)].$$

Moreover, every quadratic operator $\mathbf{Q} : \mathbb{R}^m \to \mathbb{R}^n$ can be written in the coordinate form as follows

$$\mathbf{Q}(x) = \left(\sum_{j,k=1}^{m} q_{1jk} x_j x_k, \sum_{j,k=1}^{m} q_{2jk} x_j x_k, \cdots, \sum_{j,k=1}^{m} q_{njk} x_j x_k \right)^T, \tag{2.3}$$

where $Q = (q_{ijk})_{i,j,k=1}^{n,m,m}$ is a hyper-matrix such that $q_{ijk} = q_{ikj}$ for all $i = \overline{1,n}$ and $j, k = \overline{1,m}$.

If we define vectors $\mathsf{Q}_{jk} = (q_{1jk}, q_{2jk} \dots, q_{njk})^T \in \mathbb{R}^n$, $j, k = \overline{1,m}$ then the quadratic operator (2.3) takes the following form

$$\mathbf{Q}(x) = \sum_{j,k=1}^{m} \mathsf{Q}_{jk} x_j x_k. \tag{2.4}$$

Let $\mathbf{Q}\oplus\mathbf{L}:\mathbb{R}^m\to\mathbb{R}^n$ be the sum of quadratic and linear operators

$$\mathbf{Q}\oplus\mathbf{L}(x)=\mathbf{Q}(x)+\mathbf{L}(x). \tag{2.5}$$

Problem 2.1 (Nonlinear Preserver Problem). *Describe all quadratic plus linear operators* $\mathbf{Q}\oplus\mathbf{L}:\mathbb{R}^m\to\mathbb{R}^n$ *which preserve the pure states of the quantum systems, i.e.,*

$$\mathbf{Q}\oplus\mathbf{L}(\mathcal{S}^{m-1})\subset\mathcal{S}^{n-1} \tag{2.6}$$

3. Quadratic plus Linear operators

In this section we would like to solve NPP problem on small dimensional spaces.

3.1. Two dimensional space

In this section, we shall describe all quadratic plus linear operators $\mathbf{Q}\oplus\mathbf{L}:\mathbb{R}^2\to\mathbb{R}^n$ in which one has that

$$\mathbf{Q}\oplus\mathbf{L}(\mathcal{S}^1)\subset\mathcal{S}^{n-1} \tag{3.1}$$

The quadratic plus linear operator $\mathbf{Q}\oplus\mathbf{L}:\mathbb{R}^2\to\mathbb{R}^n$ takes the following form

$$\mathbf{Q}\oplus\mathbf{L}(x)=x_1^2\mathsf{Q}_{11}+x_1x_2\mathsf{Q}_{12}+x_2^2\mathsf{Q}_{22}+x_1\mathsf{L}_1+x_2\mathsf{L}_2 \tag{3.2}$$

where $x=(x_1,x_2)^T\in\mathbb{R}^2$ and $\mathsf{Q}_{11},\mathsf{Q}_{12},\mathsf{Q}_{22},\mathsf{L}_1$, and L_2 are vectors taken from $\mathbb{R}^n$.

Remark 3.1. *It is worth mentioning that the quadratic plus linear operator* $\mathbf{Q}\oplus\mathbf{L}:\mathbb{R}^2\to\mathbb{R}^n$ *should actually take the following form*

$$\mathbf{Q}\oplus\mathbf{L}(x)=x_1^2\mathsf{Q}_{11}+2x_1x_2\mathsf{Q}_{12}+x_2^2\mathsf{Q}_{22}+x_1\mathsf{L}_1+x_2\mathsf{L}_2$$

However, by abusing notations, the coefficient Q_{12} *in* (3.2) *should be understood as* $2\mathsf{Q}_{12}$.

Theorem 3.2. *Let* $\mathbf{Q}\oplus\mathbf{L}:\mathbb{R}^2\to\mathbb{R}^n$ *be a quadratic plus linear operator given by* (3.2). *Then, one has that* $\mathbf{Q}\oplus\mathbf{L}(\mathcal{S}^1)\subset\mathcal{S}^{n-1}$ *if and only if the following conditions hold true:*

- $\|\mathsf{Q}_{11}\|^2+\|\mathsf{L}_1\|^2=\|\mathsf{Q}_{22}\|^2+\|\mathsf{L}_2\|^2=1,\ \ \mathsf{Q}_{11}\perp\mathsf{L}_1\ \ \mathsf{Q}_{22}\perp\mathsf{L}_2;$
- $\|\mathsf{Q}_{11}-\mathsf{Q}_{22}\|=\|\mathsf{Q}_{12}\|;$
- $(\mathsf{Q}_{11},\mathsf{Q}_{12})=(\mathsf{Q}_{22},\mathsf{Q}_{12})=-(\mathsf{L}_1,\mathsf{L}_2);$
- $(\mathsf{Q}_{11},\mathsf{L}_2)+(\mathsf{Q}_{12},\mathsf{L}_1)=(\mathsf{Q}_{22},\mathsf{L}_1)+(\mathsf{Q}_{12},\mathsf{L}_2)=0.$

Proof. It is clear that one has $\mathbf{Q}\oplus\mathbf{L}(\mathcal{S}^1)\subset\mathcal{S}^{n-1}$ if and only if $\|\mathbf{Q}(x)+\mathbf{L}(x)\|=1$ for any $x\in\mathcal{S}^1$. Since $-x,x\in\mathcal{S}^1$, we get from the last equality that $\|\mathbf{Q}(x)+\mathbf{L}(x)\|=1$ and $\|\mathbf{Q}(-x)+\mathbf{L}(-x)\|=\|\mathbf{Q}(x)-\mathbf{L}(x)\|=1$ for any $x\in\mathcal{S}^1$. Therefore, $\mathbf{Q}\oplus\mathbf{L}(\mathcal{S}^1)\subset\mathcal{S}^{n-1}$ is true if and only if $\|\mathbf{Q}(x)\|^2+\|\mathbf{L}(x)\|^2=1$ and $\mathbf{Q}(x)\perp\mathbf{L}(x)$ for any $x\in\mathcal{S}^1$.

We know that $\mathbf{Q}(e_1)=\mathsf{Q}_{11}$, $\mathbf{Q}(e_2)=\mathsf{Q}_{22}$ and $\mathbf{L}(e_1)=\mathsf{L}_1$, $\mathbf{L}(e_2)=\mathsf{L}_2$.

Let us describe all quadratic and linear operators such that $\mathbf{Q}(x)\perp\mathbf{L}(x)$ or $(\mathbf{Q}(x),\mathbf{L}(x))=0$ for any $x\in\mathcal{S}^1$.

Let $\mu_1,\mu_2\in\mathbb{R}$ be any two numbers such that $\mu_1^2+\mu_2^2\neq0$. We choose $x_1=\dfrac{\mu_1}{\sqrt{\mu_1^2+\mu_2^2}}$ and $x_2=\dfrac{\mu_2}{\sqrt{\mu_1^2+\mu_2^2}}$. It is clear that $\mu=(\mu_1,\mu_2)^T\in\mathbb{R}^2$ and $x=\dfrac{\mu}{\|\mu\|}\in\mathcal{S}^1$. Then $(\mathbf{Q}(x),\mathbf{L}(x))=0$ for any $x\in\mathcal{S}^1$ if and only if $\big(\mathbf{Q}(\mu),\mathbf{L}(\mu)\big)=0$ for any $\mu\in\mathbb{R}^2$. This means that

$$\begin{aligned}0&=\Big(\mu_1^2\mathsf{Q}_{11}+\mu_1\mu_2\mathsf{Q}_{12}+\mu_2^2\mathsf{Q}_{22}\ ,\ \mu_1\mathsf{L}_1+\mu_2\mathsf{L}_2\Big)\\&=\mu_1^3(\mathsf{Q}_{11},\mathsf{L}_1)+\mu_1^2\mu_2\Big[(\mathsf{Q}_{11},\mathsf{L}_2)+(\mathsf{Q}_{12},\mathsf{L}_1)\Big]+\mu_1\mu_2^2\Big[(\mathsf{Q}_{22},\mathsf{L}_1)+(\mathsf{Q}_{12},\mathsf{L}_2)\Big]+\mu_2^3(\mathsf{Q}_{22},\mathsf{L}_2)\end{aligned}$$

for any $\mu_1, \mu_2 \in \mathbb{R}$. Hence, we get that

$$(\mathsf{Q}_{11}, \mathsf{L}_1) = (\mathsf{Q}_{22}, \mathsf{L}_2) = 0, \quad (\mathsf{Q}_{11}, \mathsf{L}_2) + (\mathsf{Q}_{12}, \mathsf{L}_1) = (\mathsf{Q}_{22}, \mathsf{L}_1) + (\mathsf{Q}_{12}, \mathsf{L}_2) = 0.$$

Let us now describe all quadratic and linear operators such that $\|\mathbf{Q}(x)\|^2 + \|\mathbf{L}(x)\|^2 = 1$ for any $x \in \mathcal{S}^1$. Since $\|x\| = 1$, we have that $\|\mathbf{Q}(x)\|^2 + \|\mathbf{L}(x)\|^2 = 1$ for any $x \in \mathcal{S}^1$ if and only if $\|\mathbf{Q}(x)\|^2 + \|\mathbf{L}(x)\|^2 \cdot \|x\|^2 = \|x\|^4$ for any $x \in \mathcal{S}^1$.

Due to notations given above, we choose any nonzero element $\mu = (\mu_1, \mu_2)^T \in \mathbb{R}^2$. Then $x = \dfrac{\mu}{\|\mu\|} \in \mathcal{S}^1$. Consequently, we have that $\|\mathbf{Q}(x)\|^2 + \|\mathbf{L}(x)\|^2 \cdot \|x\|^2 = \|x\|^4$ for any $x \in \mathcal{S}^1$ if and only if

$$\|\mathbf{Q}(\mu)\|^2 + \|\mathbf{L}(\mu)\|^2 \cdot \|\mu\|^2 = \|\mu\|^4, \quad \forall\ \mu \in \mathbb{R}^2.$$

We know that

$$\begin{aligned}
\|\mathbf{Q}(\mu)\|^2 &= \mu_1^4(\mathsf{Q}_{11}, \mathsf{Q}_{11}) + \mu_1^2\mu_2^2\Big[(\mathsf{Q}_{12}, \mathsf{Q}_{12}) + 2(\mathsf{Q}_{11}, \mathsf{Q}_{22})\Big] + \mu_2^4(\mathsf{Q}_{22}, \mathsf{Q}_{22}) \\
&\quad +2\mu_1^3\mu_2(\mathsf{Q}_{11}, \mathsf{Q}_{12}) + 2\mu_1\mu_2^3(\mathsf{Q}_{22}, \mathsf{Q}_{12}), \\
\|\mathbf{L}(\mu)\|^2 &= \mu_1^2(\mathsf{L}_1, \mathsf{L}_1) + 2\mu_1\mu_2(\mathsf{L}_1, \mathsf{L}_2) + \mu_2^2(\mathsf{L}_2, \mathsf{L}_2), \\
\|\mu\|^2 &= \mu_1^2 + \mu_2^2, \quad \|\mu\|^4 = \mu_1^4 + 2\mu_1^2\mu_2^2 + \mu_2^4.
\end{aligned}$$

Therefore, we obtain that

$$\begin{aligned}
&\mu_1^4\Big[(\mathsf{Q}_{11}, \mathsf{Q}_{11}) + (\mathsf{L}_1, \mathsf{L}_1) - 1\Big] + \mu_1^2\mu_2^2\Big[(\mathsf{Q}_{12}, \mathsf{Q}_{12}) + 2(\mathsf{Q}_{11}, \mathsf{Q}_{22}) + (\mathsf{L}_1, \mathsf{L}_1) + (\mathsf{L}_2, \mathsf{L}_2) - 2\Big] \\
&\quad +\mu_2^4\Big[(\mathsf{Q}_{22}, \mathsf{Q}_{22}) + (\mathsf{L}_2, \mathsf{L}_2) - 1\Big] + 2\mu_1^3\mu_2\Big[(\mathsf{Q}_{11}, \mathsf{Q}_{12}) + (\mathsf{L}_1, \mathsf{L}_2)\Big] \\
&\quad +2\mu_1\mu_2^3\Big[(\mathsf{Q}_{22}, \mathsf{Q}_{12}) + (\mathsf{L}_1, \mathsf{L}_2)\Big] = 0
\end{aligned}$$

for any $\mu_1, \mu_2 \in \mathbb{R}$. Consequently, we get that

$$\begin{aligned}
(\mathsf{Q}_{11}, \mathsf{Q}_{11}) + (\mathsf{L}_1, \mathsf{L}_1) = (\mathsf{Q}_{22}, \mathsf{Q}_{22}) + (\mathsf{L}_2, \mathsf{L}_2) &= 1, \\
(\mathsf{Q}_{12}, \mathsf{Q}_{12}) + 2(\mathsf{Q}_{11}, \mathsf{Q}_{22}) + (\mathsf{L}_1, \mathsf{L}_1) + (\mathsf{L}_2, \mathsf{L}_2) - 2 &= \\
(\mathsf{Q}_{12}, \mathsf{Q}_{12}) + 2(\mathsf{Q}_{11}, \mathsf{Q}_{22}) - (\mathsf{Q}_{11}, \mathsf{Q}_{11}) - (\mathsf{Q}_{22}, \mathsf{Q}_{22}) &= 0, \\
(\mathsf{Q}_{11}, \mathsf{Q}_{12}) + (\mathsf{L}_1, \mathsf{L}_2) = (\mathsf{Q}_{22}, \mathsf{Q}_{12}) + (\mathsf{L}_1, \mathsf{L}_2) &= 0.
\end{aligned}$$

This completes the proof. □

Let $\mathbf{Q} : \mathbb{R}^2 \to \mathbb{R}^n$ be a quadratic operator

$$\mathbf{Q}(x) = x_1^2\mathsf{Q}_{11} + x_1x_2\mathsf{Q}_{12} + x_2^2\mathsf{Q}_{22}, \tag{3.3}$$

where $x = (x_1, x_2)^T \in \mathbb{R}^2$ and $\mathsf{Q}_{11}, \mathsf{Q}_{12}$, and Q_{22} are vectors taken from $\mathbb{R}^n$. We can immediately get the following corollary from Theorem 3.2.

Corollary 3.3. *Let* $\mathbf{Q} : \mathbb{R}^2 \to \mathbb{R}^n$ *be a quadratic operator given by* (3.3). *Then, one has that* $\mathbf{Q}(\mathcal{S}^1) \subset \mathcal{S}^{n-1}$ *if and only if the following conditions hold true:*

- $\|\mathsf{Q}_{11}\| = \|\mathsf{Q}_{22}\| = 1$, $\|\mathsf{Q}_{11} - \mathsf{Q}_{22}\| = \|\mathsf{Q}_{12}\|$;
- $\mathsf{Q}_{11} \perp \mathsf{Q}_{12}$, $\mathsf{Q}_{22} \perp \mathsf{Q}_{12}$;

We may describe all quadratic operator $\mathbf{Q} : \mathbb{R}^2 \to \mathbb{R}^2$ in which one has that $\mathbf{Q}(\mathcal{S}^1) \subset \mathcal{S}^1$.

Theorem 3.4. *Let* $\mathbf{Q}:\mathbb{R}^2\to\mathbb{R}^2$ *be a quadratic operator given by* (3.3). *Then, one has that* $\mathbf{Q}(\mathcal{S}^1)\subset\mathcal{S}^1$ *if and only if the quadratic operator* $\mathbf{Q}:\mathbb{R}^2\to\mathbb{R}^2$ *has either one of the following forms:*

(i) There are two perpendicular vectors $\mathbf{a},\mathbf{b}\in\mathcal{S}^1$ *such that* $\mathbf{Q}_{\mathbf{a},\mathbf{b}}(x)=(x_1^2-x_2^2)\mathbf{a}+2x_1x_2\mathbf{b}$;

(ii) There is a vector $\mathbf{a}\in\mathcal{S}^1$ *such that* $\mathbf{Q}_{\mathbf{a}}(x)=\mathbf{a}$.

Proof. Let $\mathbf{Q}:\mathbb{R}^2\to\mathbb{R}^2$ be a quadratic operator

$$\mathbf{Q}(x)=x_1^2\mathrm{Q}_{11}+x_1x_2\mathrm{Q}_{12}+x_2^2\mathrm{Q}_{22},$$

where $\mathrm{Q}_{11},\mathrm{Q}_{12}$, and Q_{22} are vectors taken from $\mathbb{R}^2$. Let us consider two cases.

CASE I. Let $\mathrm{Q}_{11}=\mathrm{Q}_{22}=\mathbf{a}$. Then due to Corollary 3.3, one has that $\mathbf{a}\in\mathcal{S}^1$ and $\mathrm{Q}_{12}=(0,0)$. Thus, the quadratic operator takes the form $\mathbf{Q}_{\mathbf{a}}(x)=(x_1^2+x_2^2)\mathbf{a}=\mathbf{a}$.

CASE II. Let $\mathrm{Q}_{11}\neq\mathrm{Q}_{22}$. Then due to Corollary 3.3, one has that $\mathrm{Q}_{11}\perp\mathrm{Q}_{12}$ and $\mathrm{Q}_{22}\perp\mathrm{Q}_{12}$. Since $\mathrm{Q}_{11},\mathrm{Q}_{12}$, and Q_{22} are nonzero vectors of $\mathbb{R}^2$, vectors $\mathrm{Q}_{11},\mathrm{Q}_{22}$ should be linearly dependent. Since $\|\mathrm{Q}_{11}\|=\|\mathrm{Q}_{22}\|=1$, and $\mathrm{Q}_{11}\neq\mathrm{Q}_{22}$, we get that $\mathrm{Q}_{11}=-\mathrm{Q}_{22}$. Consequently, we have that $\|\mathrm{Q}_{12}\|=\|\mathrm{Q}_{11}-\mathrm{Q}_{22}\|=2\|\mathrm{Q}_{11}\|=2$ and $\mathrm{Q}_{11}\perp\mathrm{Q}_{12}$. Let $\mathrm{Q}_{11}=\mathbf{a}$ and $\frac{1}{2}\mathrm{Q}_{12}=\mathbf{b}$. It is clear that $\mathbf{a},\mathbf{b}\in\mathcal{S}^1$ and $\mathbf{a}\perp\mathbf{b}$. Then, we get that $\mathbf{Q}_{\mathbf{a},\mathbf{b}}(x)=(x_1^2-x_2^2)\mathbf{a}+2x_1x_2\mathbf{b}$ for any $x\in\mathcal{S}^1$. This completes the proof.

□

Let us discuss the dynamics of the quadratic mapping $\mathbf{Q}_{\mathbf{a},\mathbf{b}}:\mathbb{R}^2\to\mathbb{R}^2$

$$\mathbf{Q}_{\mathbf{a},\mathbf{b}}(x)=(x_1^2-x_2^2)\mathbf{a}+2x_1x_2\mathbf{b} \tag{3.4}$$

where $\mathbf{a},\mathbf{b}\in\mathbb{R}^2$ are any two fixed vectors such that $\|\mathbf{a}\|=\|\mathbf{b}\|=1$ and $\mathbf{a}\perp\mathbf{b}$.

The dynamics of the quadratic mapping (3.4) outside of the unit circle $\mathcal{S}^1$ is trivial. Namely, if $\|x\|<1$ then $\|\mathbf{Q}_{\mathbf{a},\mathbf{b}}{}^{(n)}(x)\|=\|x\|^{2^n}\to0$ and if $\|x\|>1$ then $\|\mathbf{Q}_{\mathbf{a},\mathbf{b}}{}^{(n)}(x)\|=\|x\|^{2^n}\to\infty$. Therefore, it is enough to study the dynamics of the quadratic mapping (3.4) on the invariant unit circle $\mathcal{S}^1$.

Since $\|\mathbf{a}\|=\|\mathbf{b}\|=1$ and $\mathbf{a}\perp\mathbf{b}$, we may assume that $\mathbf{a}=(\lambda_1,\lambda_2)$ and $\mathbf{b}=(-\lambda_2,\lambda_1)$. In this case, the quadratic operator (3.4) takes the following form:

$$\mathbf{Q}_{\mathbf{a},\mathbf{b}}(x)=\Big((x_1^2-x_2^2)\lambda_1-2x_1x_2\lambda_2,(x_1^2-x_2^2)\lambda_2+2x_1x_2\lambda_1\Big)$$

Let $z=x_1+ix_2$ and $\lambda=\lambda_1+i\lambda_2$ be complex numbers. Let us consider the complex quadratic function $f_\lambda:\mathbb{C}\to\mathbb{C}$ where $f_\lambda(z)=\lambda z^2$. It is easy to check that

$$f_\lambda(z)=(x_1^2-x_2^2)\lambda_1-2x_1x_2\lambda_2+i((x_1^2-x_2^2)\lambda_2+2x_1x_2\lambda_1).$$

Consequently, the quadratic operator $\mathbf{Q}_{\mathbf{a},\mathbf{b}}:\mathcal{S}^1\to\mathcal{S}^1$ can be considered as the complex quadratic function $f_\lambda:\mathbb{C}\to\mathbb{C}$, i.e.,

$$\mathbf{Q}_{\mathbf{a},\mathbf{b}}(\cdot)=\Big(Re(f_\lambda(\cdot)),Im(f_\lambda(\cdot))\Big).$$

The dynamics of the complex quadratic function $f_\lambda(z)=\lambda z^2$ on the unit circle, where $|\lambda|=1$, was studied very well (for example, see [2]). Based on this study, we may describe the dynamics of the quadratic operator (3.4).

Theorem 3.5. *Let* $\mathbf{Q}_{\mathbf{a},\mathbf{b}}:\mathcal{S}^1\to\mathcal{S}^1$ *be the quadratic operator given by* (3.4). *Then* $\mathbf{Q}_{\mathbf{a},\mathbf{b}}:\mathcal{S}^1\to\mathcal{S}^1$ *is chaotic in the sense of Devaney. Namely, the following statements hold true:*

(i) The set of periodic points are dense in the circle;

(ii) Every arc contains infinitely many points in which its trajectory is dense in the circle;

(iii) Every arc contains infinitely many points which are eventually fixed points.

3.2. Three dimensional space

In this section, we shall describe all quadratic plus linear operators $\mathbf{Q} \oplus \mathbf{L} : \mathbb{R}^3 \to \mathbb{R}^n$ in which one has that

$$\mathbf{Q} \oplus \mathbf{L}(\mathcal{S}^2) \subset \mathcal{S}^{n-1} \tag{3.5}$$

The quadratic plus linear operator $\mathbf{Q} \oplus \mathbf{L} : \mathbb{R}^3 \to \mathbb{R}^n$ takes the following form

$$\mathbf{Q} \oplus \mathbf{L}(x) = x_1^2\mathsf{Q}_{11} + x_2^2\mathsf{Q}_{22} + x_3^2\mathsf{Q}_{33} + x_1x_2\mathsf{Q}_{12} + x_1x_3\mathsf{Q}_{13} + x_2x_3\mathsf{Q}_{23} + x_1\mathsf{L}_1 + x_2\mathsf{L}_2 + x_3\mathsf{L}_3 \tag{3.6}$$

where $x = (x_1, x_2, x_3)^T \in \mathbb{R}^3$ and $\mathsf{Q}_{11}, \mathsf{Q}_{22}, \mathsf{Q}_{33}, \mathsf{Q}_{12}, \mathsf{Q}_{13}, \mathsf{Q}_{23}, \mathsf{L}_1, \mathsf{L}_2, \mathsf{L}_3 \in \mathbb{R}^n$.

Remark 3.6. *It is worth mentioning that the quadratic plus linear operator $\mathbf{Q} \oplus \mathbf{L} : \mathbb{R}^3 \to \mathbb{R}^n$ should actually take the following form*

$$\mathbf{Q} \oplus \mathbf{L}(x) = x_1^2\mathsf{Q}_{11} + x_2^2\mathsf{Q}_{22} + x_3^2\mathsf{Q}_{33} + 2x_1x_2\mathsf{Q}_{12} + 2x_1x_3\mathsf{Q}_{13} + 2x_2x_3\mathsf{Q}_{23} + x_1\mathsf{L}_1 + x_2\mathsf{L}_2 + x_3\mathsf{L}_3$$

However, by abusing notations, the coefficients $\mathsf{Q}_{12}, \mathsf{Q}_{13}, \mathsf{Q}_{23}$ in (3.6) *should be understood as* $2\mathsf{Q}_{12}, 2\mathsf{Q}_{13}, 2\mathsf{Q}_{23}$.

Theorem 3.7. *Let $\mathbf{Q} \oplus \mathbf{L} : \mathbb{R}^3 \to \mathbb{R}^n$ be a quadratic plus linear operator given by* (3.6)*. Then, one has that $\mathbf{Q} \oplus \mathbf{L}(\mathcal{S}^2) \subset \mathcal{S}^{n-1}$ if and only if the following conditions hold true:*

- $\|\mathsf{Q}_{11}\|^2 + \|\mathsf{L}_1\|^2 = \|\mathsf{Q}_{22}\|^2 + \|\mathsf{L}_2\|^2 = \|\mathsf{Q}_{33}\|^2 + \|\mathsf{L}_3\|^2 = 1, \ \mathsf{Q}_{11} \perp \mathsf{L}_1 \ \ \mathsf{Q}_{22} \perp \mathsf{L}_2, \ \mathsf{Q}_{33} \perp \mathsf{L}_3;$
- $\|\mathsf{Q}_{11} - \mathsf{Q}_{22}\| = \|\mathsf{Q}_{12}\|, \ \|\mathsf{Q}_{11} - \mathsf{Q}_{33}\| = \|\mathsf{Q}_{13}\|, \ \|\mathsf{Q}_{22} - \mathsf{Q}_{33}\| = \|\mathsf{Q}_{23}\|;$
- $(\mathsf{Q}_{11}, \mathsf{Q}_{12}) = (\mathsf{Q}_{22}, \mathsf{Q}_{12}) = -(\mathsf{L}_1, \mathsf{L}_2) = (\mathsf{Q}_{33}, \mathsf{Q}_{12}) + (\mathsf{Q}_{13}, \mathsf{Q}_{23}),$
  $(\mathsf{Q}_{11}, \mathsf{Q}_{13}) = (\mathsf{Q}_{33}, \mathsf{Q}_{13}) = -(\mathsf{L}_1, \mathsf{L}_3) = (\mathsf{Q}_{22}, \mathsf{Q}_{13}) + (\mathsf{Q}_{12}, \mathsf{Q}_{23}),$
  $(\mathsf{Q}_{22}, \mathsf{Q}_{23}) = (\mathsf{Q}_{33}, \mathsf{Q}_{23}) = -(\mathsf{L}_2, \mathsf{L}_3) = (\mathsf{Q}_{11}, \mathsf{Q}_{23}) + (\mathsf{Q}_{12}, \mathsf{Q}_{13});$
- $(\mathsf{Q}_{11}, \mathsf{L}_2) + (\mathsf{Q}_{12}, \mathsf{L}_1) = (\mathsf{Q}_{22}, \mathsf{L}_1) + (\mathsf{Q}_{12}, \mathsf{L}_2) = 0,$
  $(\mathsf{Q}_{11}, \mathsf{L}_3) + (\mathsf{Q}_{13}, \mathsf{L}_1) = (\mathsf{Q}_{33}, \mathsf{L}_1) + (\mathsf{Q}_{13}, \mathsf{L}_3) = 0,$
  $(\mathsf{Q}_{22}, \mathsf{L}_3) + (\mathsf{Q}_{23}, \mathsf{L}_2) = (\mathsf{Q}_{33}, \mathsf{L}_2) + (\mathsf{Q}_{23}, \mathsf{L}_3) = 0;$
- $(\mathsf{Q}_{12}, \mathsf{L}_3) + (\mathsf{Q}_{13}, \mathsf{L}_2) + (\mathsf{Q}_{23}, \mathsf{L}_1) = 0.$

Proof. It is clear that one has $\mathbf{Q} \oplus \mathbf{L}(\mathcal{S}^2) \subset \mathcal{S}^{n-1}$ if and only if $\|\mathbf{Q}(x) + \mathbf{L}(x)\| = 1$ for any $x \in \mathcal{S}^2$. Since $-x, x \in \mathcal{S}^2$, we get from the last equality that $\|\mathbf{Q}(x) + \mathbf{L}(x)\| = 1$ and $\|\mathbf{Q}(-x) + \mathbf{L}(-x)\| = \|\mathbf{Q}(x) - \mathbf{L}(x)\| = 1$ for any $x \in \mathcal{S}^2$. Therefore, $\mathbf{Q} \oplus \mathbf{L}(\mathcal{S}^2) \subset \mathcal{S}^{n-1}$ is true if and only if $\|\mathbf{Q}(x)\|^2 + \|\mathbf{L}(x)\|^2 = 1$ and $\mathbf{Q}(x) \perp \mathbf{L}(x)$ for any $x \in \mathcal{S}^2$.

We know that $\mathbf{Q}(e_1) = \mathsf{Q}_{11}$, $\mathbf{Q}(e_2) = \mathsf{Q}_{22}$, $\mathbf{Q}(e_3) = \mathsf{Q}_{33}$, $\mathbf{L}(e_1) = \mathsf{L}_1$, $\mathbf{L}(e_2) = \mathsf{L}_2$, $\mathbf{L}(e_3) = \mathsf{L}_3$.

Let us describe all quadratic and linear operators such that $\mathbf{Q}(x) \perp \mathbf{L}(x)$ or $(\mathbf{Q}(x), \mathbf{L}(x)) = 0$ for any $x \in \mathcal{S}^2$.

Let $\mu_1, \mu_2, \mu_3 \in \mathbb{R}$ be any three numbers such that $\mu_1^2 + \mu_2^2 + \mu_3^2 \neq 0$. We choose $x_1 = \dfrac{\mu_1}{\sqrt{\mu_1^2 + \mu_2^2 + \mu_3^2}}$, $x_2 = \dfrac{\mu_2}{\sqrt{\mu_1^2 + \mu_2^2 + \mu_3^2}}$ and $x_3 = \dfrac{\mu_3}{\sqrt{\mu_1^2 + \mu_2^2 + \mu_3^2}}$. It is clear that $\mu = (\mu_1, \mu_2, \mu_3)^T \in \mathbb{R}^3$ and $x = \dfrac{\mu}{\|\mu\|} \in \mathcal{S}^2$. Then $(\mathbf{Q}(x), \mathbf{L}(x)) = 0$ for any $x \in \mathcal{S}^2$ if and only if $\big(\mathbf{Q}(\mu), \mathbf{L}(\mu)\big) = 0$ for any $\mu \in \mathbb{R}^3$. This means that

$$\begin{aligned} 0 &= \Big(\mu_1^2\mathsf{Q}_{11} + \mu_2^2\mathsf{Q}_{22} + \mu_3^2\mathsf{Q}_{33} + \mu_1\mu_2\mathsf{Q}_{12} + \mu_1\mu_3\mathsf{Q}_{13} + \mu_2\mu_3\mathsf{Q}_{23}\ ,\ \mu_1\mathsf{L}_1 + \mu_2\mathsf{L}_2 + \mu_3\mathsf{L}_3\Big) \\ &= \mu_1^3(\mathsf{Q}_{11}, \mathsf{L}_1) + \mu_2^3(\mathsf{Q}_{22}, \mathsf{L}_2) + \mu_3^3(\mathsf{Q}_{33}, \mathsf{L}_3) \\ &\quad +\mu_1^2\mu_2\Big[(\mathsf{Q}_{11}, \mathsf{L}_2) + (\mathsf{Q}_{12}, \mathsf{L}_1)\Big] + \mu_1\mu_2^2\Big[(\mathsf{Q}_{22}, \mathsf{L}_1) + (\mathsf{Q}_{12}, \mathsf{L}_2)\Big] \\ &\quad +\mu_1^2\mu_3\Big[(\mathsf{Q}_{11}, \mathsf{L}_3) + (\mathsf{Q}_{13}, \mathsf{L}_1)\Big] + \mu_1\mu_3^2\Big[(\mathsf{Q}_{33}, \mathsf{L}_1) + (\mathsf{Q}_{13}, \mathsf{L}_3)\Big] \\ &\quad +\mu_2^2\mu_3\Big[(\mathsf{Q}_{22}, \mathsf{L}_3) + (\mathsf{Q}_{23}, \mathsf{L}_2)\Big] + \mu_2\mu_3^2\Big[(\mathsf{Q}_{33}, \mathsf{L}_2) + (\mathsf{Q}_{23}, \mathsf{L}_3)\Big] \\ &\quad +\mu_1\mu_2\mu_3\Big[(\mathsf{Q}_{12}, \mathsf{L}_3) + (\mathsf{Q}_{13}, \mathsf{L}_2) + (\mathsf{Q}_{23}, \mathsf{L}_1)\Big] \end{aligned}$$

for any $\mu_1, \mu_2, \mu_3 \in \mathbb{R}$. Hence, we get that

$$\begin{aligned} (\mathsf{Q}_{11}, \mathsf{L}_1) = (\mathsf{Q}_{22}, \mathsf{L}_2) = (\mathsf{Q}_{33}, \mathsf{L}_3) = 0, \\ (\mathsf{Q}_{11}, \mathsf{L}_2) + (\mathsf{Q}_{12}, \mathsf{L}_1) = (\mathsf{Q}_{22}, \mathsf{L}_1) + (\mathsf{Q}_{12}, \mathsf{L}_2) = 0, \\ (\mathsf{Q}_{11}, \mathsf{L}_3) + (\mathsf{Q}_{13}, \mathsf{L}_1) = (\mathsf{Q}_{33}, \mathsf{L}_1) + (\mathsf{Q}_{13}, \mathsf{L}_3) = 0, \\ (\mathsf{Q}_{22}, \mathsf{L}_3) + (\mathsf{Q}_{23}, \mathsf{L}_2) = (\mathsf{Q}_{33}, \mathsf{L}_2) + (\mathsf{Q}_{23}, \mathsf{L}_3) = 0, \\ (\mathsf{Q}_{12}, \mathsf{L}_3) + (\mathsf{Q}_{13}, \mathsf{L}_2) + (\mathsf{Q}_{23}, \mathsf{L}_1) = 0. \end{aligned}$$

Let us now describe all quadratic and linear operators such that $\|\mathbf{Q}(x)\|^2 + \|\mathbf{L}(x)\|^2 = 1$ for any $x \in \mathcal{S}^2$. Since $\|x\| = 1$, we have that $\|\mathbf{Q}(x)\|^2 + \|\mathbf{L}(x)\|^2 = 1$ for any $x \in \mathcal{S}^2$ if and only if $\|\mathbf{Q}(x)\|^2 + \|\mathbf{L}(x)\|^2 \cdot \|x\|^2 = \|x\|^4$ for any $x \in \mathcal{S}^2$.

Due to notations given above, we choose any nonzero element $\mu = (\mu_1, \mu_2, \mu_3)^T \in \mathbb{R}^3$. Then $x = \dfrac{\mu}{\|\mu\|} \in \mathcal{S}^2$. Consequently, we have that $\|\mathbf{Q}(x)\|^2 + \|\mathbf{L}(x)\|^2 \cdot \|x\|^2 = \|x\|^4$ for any $x \in \mathcal{S}^2$ if and only if

$$\|\mathbf{Q}(\mu)\|^2 + \|\mathbf{L}(\mu)\|^2 \cdot \|\mu\|^2 = \|\mu\|^4, \quad \forall\ \mu \in \mathbb{R}^3.$$

We know that

$$\begin{aligned} \|\mathbf{Q}(\mu)\|^2 &= \mu_1^4(\mathsf{Q}_{11}, \mathsf{Q}_{11}) + \mu_2^4(\mathsf{Q}_{22}, \mathsf{Q}_{22}) + \mu_3^4(\mathsf{Q}_{33}, \mathsf{Q}_{33}) \\ &\quad +\mu_1^2\mu_2^2\Big[(\mathsf{Q}_{12}, \mathsf{Q}_{12}) + 2(\mathsf{Q}_{11}, \mathsf{Q}_{22})\Big] + \mu_1^2\mu_3^2\Big[(\mathsf{Q}_{13}, \mathsf{Q}_{13}) + 2(\mathsf{Q}_{11}, \mathsf{Q}_{33})\Big] \\ &\quad +\mu_2^2\mu_3^2\Big[(\mathsf{Q}_{23}, \mathsf{Q}_{23}) + 2(\mathsf{Q}_{22}, \mathsf{Q}_{33})\Big] \\ &\quad +2\mu_1^3\mu_2(\mathsf{Q}_{11}, \mathsf{Q}_{12}) + 2\mu_1\mu_2^3(\mathsf{Q}_{22}, \mathsf{Q}_{12}) + 2\mu_1^3\mu_3(\mathsf{Q}_{11}, \mathsf{Q}_{13}) + 2\mu_1\mu_3^3(\mathsf{Q}_{33}, \mathsf{Q}_{13}) \\ &\quad +2\mu_2^3\mu_3(\mathsf{Q}_{22}, \mathsf{Q}_{23}) + 2\mu_2\mu_3^3(\mathsf{Q}_{33}, \mathsf{Q}_{23}) \\ &\quad +2\mu_1^2\mu_2\mu_3\Big[(\mathsf{Q}_{11}, \mathsf{Q}_{23}) + (\mathsf{Q}_{12}, \mathsf{Q}_{13})\Big] + 2\mu_1\mu_2^2\mu_3\Big[(\mathsf{Q}_{22}, \mathsf{Q}_{13}) + (\mathsf{Q}_{12}, \mathsf{Q}_{23})\Big] \\ &\quad +2\mu_1\mu_2\mu_3^2\Big[(\mathsf{Q}_{33}, \mathsf{Q}_{12}) + (\mathsf{Q}_{13}, \mathsf{Q}_{23})\Big], \\ \|\mathbf{L}(\mu)\|^2 &= \mu_1^2(\mathsf{L}_1, \mathsf{L}_1) + \mu_2^2(\mathsf{L}_2, \mathsf{L}_2) + \mu_3^2(\mathsf{L}_3, \mathsf{L}_3) \\ &\quad +2\mu_1\mu_2(\mathsf{L}_1, \mathsf{L}_2) + 2\mu_1\mu_3(\mathsf{L}_1, \mathsf{L}_3) + 2\mu_2\mu_3(\mathsf{L}_2, \mathsf{L}_3), \\ \|\mu\|^2 &= \mu_1^2 + \mu_2^2 + \mu_3^2, \quad \|\mu\|^4 = \mu_1^4 + \mu_2^4 + \mu_3^4 + 2\mu_1^2\mu_2^2 + 2\mu_1^2\mu_3^2 + 2\mu_2^2\mu_3^2. \end{aligned}$$

Therefore, we obtain that

$$\begin{aligned}
&\mu_1^4\Big[(\mathsf{Q}_{11},\mathsf{Q}_{11})+(\mathsf{L}_1,\mathsf{L}_1)-1\Big]+\mu_2^4\Big[(\mathsf{Q}_{22},\mathsf{Q}_{22})+(\mathsf{L}_2,\mathsf{L}_2)-1\Big]+\mu_3^4\Big[(\mathsf{Q}_{33},\mathsf{Q}_{33})+(\mathsf{L}_3,\mathsf{L}_3)-1\Big]\\
&+\mu_1^2\mu_2^2\Big[(\mathsf{Q}_{12},\mathsf{Q}_{12})+2(\mathsf{Q}_{11},\mathsf{Q}_{22})+(\mathsf{L}_1,\mathsf{L}_1)+(\mathsf{L}_2,\mathsf{L}_2)-2\Big]\\
&+\mu_1^2\mu_3^2\Big[(\mathsf{Q}_{13},\mathsf{Q}_{13})+2(\mathsf{Q}_{11},\mathsf{Q}_{33})+(\mathsf{L}_1,\mathsf{L}_1)+(\mathsf{L}_3,\mathsf{L}_3)-2\Big]\\
&+\mu_2^2\mu_3^2\Big[(\mathsf{Q}_{23},\mathsf{Q}_{23})+2(\mathsf{Q}_{22},\mathsf{Q}_{33})+(\mathsf{L}_2,\mathsf{L}_2)+(\mathsf{L}_3,\mathsf{L}_3)-2\Big]\\
&+2\mu_1^3\mu_2\Big[(\mathsf{Q}_{11},\mathsf{Q}_{12})+(\mathsf{L}_1,\mathsf{L}_2)\Big]+2\mu_1\mu_2^3\Big[(\mathsf{Q}_{22},\mathsf{Q}_{12})+(\mathsf{L}_1,\mathsf{L}_2)\Big]\\
&+2\mu_1^3\mu_3\Big[(\mathsf{Q}_{11},\mathsf{Q}_{13})+(\mathsf{L}_1,\mathsf{L}_3)\Big]+2\mu_1\mu_3^3\Big[(\mathsf{Q}_{33},\mathsf{Q}_{13})+(\mathsf{L}_1,\mathsf{L}_3)\Big]\\
&+2\mu_2^3\mu_3\Big[(\mathsf{Q}_{22},\mathsf{Q}_{23})+(\mathsf{L}_2,\mathsf{L}_3)\Big]+2\mu_2\mu_3^3\Big[(\mathsf{Q}_{33},\mathsf{Q}_{23})+(\mathsf{L}_2,\mathsf{L}_3)\Big]\\
&+2\mu_1^2\mu_2\mu_3\Big[(\mathsf{Q}_{11},\mathsf{Q}_{23})+(\mathsf{Q}_{12},\mathsf{Q}_{13})+(\mathsf{L}_2,\mathsf{L}_3)\Big]+2\mu_1\mu_2^2\mu_3\Big[(\mathsf{Q}_{22},\mathsf{Q}_{13})+(\mathsf{Q}_{12},\mathsf{Q}_{23})+(\mathsf{L}_1,\mathsf{L}_3)\Big]\\
&+2\mu_1\mu_2\mu_3^2\Big[(\mathsf{Q}_{33},\mathsf{Q}_{12})+(\mathsf{Q}_{13},\mathsf{Q}_{23})+(\mathsf{L}_1,\mathsf{L}_2)\Big]=0
\end{aligned}$$

for any $\mu_1,\mu_2,\mu_3\in\mathbb{R}$. Consequently, we get that

$$\begin{aligned}
(\mathsf{Q}_{11},\mathsf{Q}_{11})+(\mathsf{L}_1,\mathsf{L}_1)=(\mathsf{Q}_{22},\mathsf{Q}_{22})+(\mathsf{L}_2,\mathsf{L}_2)=(\mathsf{Q}_{33},\mathsf{Q}_{33})+(\mathsf{L}_3,\mathsf{L}_3)&=1,\\
(\mathsf{Q}_{12},\mathsf{Q}_{12})+2(\mathsf{Q}_{11},\mathsf{Q}_{22})+(\mathsf{L}_1,\mathsf{L}_1)+(\mathsf{L}_2,\mathsf{L}_2)-2&=\\
(\mathsf{Q}_{12},\mathsf{Q}_{12})+2(\mathsf{Q}_{11},\mathsf{Q}_{22})-(\mathsf{Q}_{11},\mathsf{Q}_{11})-(\mathsf{Q}_{22},\mathsf{Q}_{22})&=0,\\
(\mathsf{Q}_{13},\mathsf{Q}_{13})+2(\mathsf{Q}_{11},\mathsf{Q}_{33})+(\mathsf{L}_1,\mathsf{L}_1)+(\mathsf{L}_3,\mathsf{L}_3)-2&=\\
(\mathsf{Q}_{13},\mathsf{Q}_{13})+2(\mathsf{Q}_{11},\mathsf{Q}_{33})-(\mathsf{Q}_{11},\mathsf{Q}_{11})-(\mathsf{Q}_{33},\mathsf{Q}_{33})&=0,\\
(\mathsf{Q}_{23},\mathsf{Q}_{23})+2(\mathsf{Q}_{22},\mathsf{Q}_{33})+(\mathsf{L}_2,\mathsf{L}_2)+(\mathsf{L}_3,\mathsf{L}_3)-2&=\\
(\mathsf{Q}_{23},\mathsf{Q}_{23})+2(\mathsf{Q}_{22},\mathsf{Q}_{33})-(\mathsf{Q}_{22},\mathsf{Q}_{22})-(\mathsf{Q}_{33},\mathsf{Q}_{33})&=0,\\
(\mathsf{Q}_{11},\mathsf{Q}_{12})+(\mathsf{L}_1,\mathsf{L}_2)=(\mathsf{Q}_{22},\mathsf{Q}_{12})+(\mathsf{L}_1,\mathsf{L}_2)&=0,\\
(\mathsf{Q}_{11},\mathsf{Q}_{13})+(\mathsf{L}_1,\mathsf{L}_3)=(\mathsf{Q}_{33},\mathsf{Q}_{13})+(\mathsf{L}_1,\mathsf{L}_3)&=0,\\
(\mathsf{Q}_{22},\mathsf{Q}_{23})+(\mathsf{L}_2,\mathsf{L}_3)=(\mathsf{Q}_{33},\mathsf{Q}_{23})+(\mathsf{L}_2,\mathsf{L}_3)&=0,\\
(\mathsf{Q}_{11},\mathsf{Q}_{23})+(\mathsf{Q}_{12},\mathsf{Q}_{13})+(\mathsf{L}_2,\mathsf{L}_3)&=0,\\
(\mathsf{Q}_{22},\mathsf{Q}_{13})+(\mathsf{Q}_{12},\mathsf{Q}_{23})+(\mathsf{L}_1,\mathsf{L}_3)&=0,\\
(\mathsf{Q}_{33},\mathsf{Q}_{12})+(\mathsf{Q}_{13},\mathsf{Q}_{23})+(\mathsf{L}_1,\mathsf{L}_2)&=0.
\end{aligned}$$

This completes the proof. □

Let $\mathbf{Q}:\mathbb{R}^3\to\mathbb{R}^n$ be a quadratic operator

$$\mathbf{Q}(x)=x_1^2\mathsf{Q}_{11}+x_2^2\mathsf{Q}_{22}+x_3^2\mathsf{Q}_{33}+x_1x_2\mathsf{Q}_{12}+x_1x_3\mathsf{Q}_{13}+x_2x_3\mathsf{Q}_{23}, \tag{3.7}$$

where $x\in\mathbb{R}^3$ and $\mathsf{Q}_{11},\mathsf{Q}_{22},\mathsf{Q}_{33},\mathsf{Q}_{12},\mathsf{Q}_{13},\mathsf{Q}_{23}\in\mathbb{R}^n$. We can immediately get the following corollary from Theorem 3.7.

Corollary 3.8. *Let $\mathbf{Q}:\mathbb{R}^3\to\mathbb{R}^n$ be a quadratic operator given by* (3.7). *Then, one has that $\mathbf{Q}(\mathcal{S}^2)\subset\mathcal{S}^{n-1}$ if and only if the following conditions hold true:*

- $\|\mathsf{Q}_{11}\|=\|\mathsf{Q}_{22}\|=\|\mathsf{Q}_{33}\|=1$;
- $\|\mathsf{Q}_{11}-\mathsf{Q}_{22}\|=\|\mathsf{Q}_{12}\|,\ \|\mathsf{Q}_{11}-\mathsf{Q}_{33}\|=\|\mathsf{Q}_{13}\|,\ \|\mathsf{Q}_{22}-\mathsf{Q}_{33}\|=\|\mathsf{Q}_{23}\|$;

- $(\mathsf{Q}_{11}, \mathsf{Q}_{12}) = (\mathsf{Q}_{22}, \mathsf{Q}_{12}) = (\mathsf{Q}_{33}, \mathsf{Q}_{12}) + (\mathsf{Q}_{13}, \mathsf{Q}_{23}) = 0,$
  $(\mathsf{Q}_{11}, \mathsf{Q}_{13}) = (\mathsf{Q}_{33}, \mathsf{Q}_{13}) = (\mathsf{Q}_{22}, \mathsf{Q}_{13}) + (\mathsf{Q}_{12}, \mathsf{Q}_{23}) = 0,$
  $(\mathsf{Q}_{22}, \mathsf{Q}_{23}) = (\mathsf{Q}_{33}, \mathsf{Q}_{23}) = (\mathsf{Q}_{11}, \mathsf{Q}_{23}) + (\mathsf{Q}_{12}, \mathsf{Q}_{13}) = 0;$

Remark 3.9. *Theorem 3.7 and Corollary 3.8 generalize results (Theorem 2 and Corollary 3) of the paper [8].*

4. Examples

In this section, we shall provide some examples for the quadratic plus linear operators on the small dimension spaces.

4.1. Two dimension space

Let $\mathbf{Q} \oplus \mathbf{L} : \mathcal{S}^1 \to \mathcal{S}^{n-1}$ be a quadratic plus linear operator

$$\mathbf{Q} \oplus \mathbf{L}(x) = x_1^2\mathsf{Q}_{11} + x_1x_2\mathsf{Q}_{12} + x_2^2\mathsf{Q}_{22} + x_1\mathsf{L}_1 + x_2\mathsf{L}_2 \tag{4.1}$$

where $x \in \mathcal{S}^1$ and $\mathsf{Q}_{11}, \mathsf{Q}_{12}, \mathsf{Q}_{22}, \mathsf{L}_1$, and L_2 are vectors in $\mathbb{R}^n$ such that they satisfies conditions of Theorem 3.2.

Let $\mathcal{L} = Span\{\mathsf{Q}_{11}, \mathsf{Q}_{12}, \mathsf{Q}_{22}, \mathsf{L}_1, \mathsf{L}_1\}$ be a linear span of the vectors $\mathsf{Q}_{11}, \mathsf{Q}_{12}, \mathsf{Q}_{22}, \mathsf{L}_1$, and L_2. One can easily see that

$$\mathbf{Q} \oplus \mathbf{L}(\mathcal{S}^1) \subset \mathcal{S}^{n-1} \cap \mathcal{L}.$$

It is clear that $\dim(\mathcal{L})$ is equal to the number of independent vectors $\mathsf{Q}_{11}, \mathsf{Q}_{12}, \mathsf{Q}_{22}, \mathsf{L}_1, \mathsf{L}_2$ and $1 \le \dim(\mathcal{L}) \le 5$. Now, we are willing to provide examples for the quadratic plus linear operators in which the dimension $\dim(\mathcal{L})$ of $\mathcal{L}$ may take any number between 1 and 5.

It is clear, for any *constant* quadratic plus linear operator, one has that $\dim(\mathcal{L}) = 1$. For any non-constant quadratic operator $\mathbf{Q} : \mathcal{S}^1 \to \mathcal{S}^1$, one has that $\dim(\mathcal{L}) = 2$ (This can be easily proven).

Example 4.1. *Let $\mathbf{Q} : \mathcal{S}^1 \to \mathcal{S}^2$ be the quadratic operator defined as follows*

$$\mathbf{Q}(x) = \left(x_1^2, \sqrt{2}x_1x_2, x_2^2\right).$$

It is clear that $\mathbf{Q}(1,0) = (1,0,0)$, $\mathbf{Q}(0,1) = (0,0,1)$, $\mathbf{Q}\left(\frac{1}{\sqrt{2}}, \frac{1}{\sqrt{2}}\right) = \left(\frac{1}{2}, \frac{1}{\sqrt{2}}, \frac{1}{2}\right)$. This means that $\mathbf{Q}(\mathcal{S}^1) \subset \mathcal{S}^2 \cap \mathcal{L} = \mathcal{S}^2$ in which $\dim(\mathcal{L}) = 3$.

Example 4.2. *Let $\mathbf{Q} \oplus \mathbf{L} : \mathcal{S}^1 \to \mathcal{S}^3$ be the quadratic plus linear operator defined as follows*

$$\mathbf{Q} \oplus \mathbf{L}(x) = \left(\frac{x_1^2 - x_2^2}{\sqrt{2}}, \sqrt{2}x_1x_2, \frac{x_1}{\sqrt{2}}, \frac{x_2}{\sqrt{2}}\right).$$

It is clear that

$$\mathbf{Q} \oplus \mathbf{L}(1,0) = \left(\frac{1}{\sqrt{2}}, 0, \frac{1}{\sqrt{2}}, 0\right), \quad \mathbf{Q} \oplus \mathbf{L}(0,1) = \left(-\frac{1}{\sqrt{2}}, 0, 0, \frac{1}{\sqrt{2}}\right),$$

$$\mathbf{Q} \oplus \mathbf{L}\left(\frac{1}{\sqrt{2}}, \frac{1}{\sqrt{2}}\right) = \left(0, \frac{1}{\sqrt{2}}, \frac{1}{2}, \frac{1}{2}\right), \quad \mathbf{Q} \oplus \mathbf{L}\left(\frac{4}{5}, \frac{3}{5}\right) = \left(\frac{7}{25\sqrt{2}}, \frac{12\sqrt{2}}{25}, \frac{4}{5\sqrt{2}}, \frac{3}{5\sqrt{2}}\right).$$

This means that $\mathbf{Q} \oplus \mathbf{L}(\mathcal{S}^1) \subset \mathcal{S}^3 \cap \mathcal{L} = \mathcal{S}^3$ in which $\dim(\mathcal{L}) = 4$.

Example 4.3. *Let* $\mathbf{Q}\oplus\mathbf{L}:\mathcal{S}^1\to\mathcal{S}^4$ *be the quadratic plus linear operator defined as follows*

$$\mathbf{Q}\oplus\mathbf{L}(x)=\left(\frac{x_1^2}{\sqrt{2}},x_1x_2,\frac{x_2^2}{\sqrt{2}},\frac{x_1}{\sqrt{2}},\frac{x_2}{\sqrt{2}}\right).$$

It is clear that

$$\mathbf{Q}\oplus\mathbf{L}(1,0)=\left(\frac{1}{\sqrt{2}},0,0,\frac{1}{\sqrt{2}},0\right),\qquad \mathbf{Q}\oplus\mathbf{L}(-1,0)=\left(\frac{1}{\sqrt{2}},0,0,-\frac{1}{\sqrt{2}},0\right),$$

$$\mathbf{Q}\oplus\mathbf{L}(0,1)=\left(0,0,\frac{1}{\sqrt{2}},0,\frac{1}{\sqrt{2}}\right),\qquad \mathbf{Q}\oplus\mathbf{L}(0,-1)=\left(0,0,\frac{1}{\sqrt{2}},0,-\frac{1}{\sqrt{2}}\right),$$

$$\mathbf{Q}\oplus\mathbf{L}\left(\frac{1}{\sqrt{2}},\frac{1}{\sqrt{2}}\right)=\left(\frac{1}{2\sqrt{2}},\frac{1}{2},\frac{1}{2\sqrt{2}},\frac{1}{2},\frac{1}{2}\right).$$

This means that $\mathbf{Q}\oplus\mathbf{L}(\mathcal{S}^1)\subset\mathcal{S}^4\cap\mathcal{L}=\mathcal{S}^4$ *in which* $\dim(\mathcal{L})=5$.

4.2. Three dimension space

Let us consider the following problem.

Let $\alpha\subset I_m=\{1,2,\cdots,m\}$ be a subset and $e_i=(\delta_{i1},\delta_{i2},\cdots,\delta_{im})^T$, $i=\overline{1,m}$ be the standard basis of $\mathbb{R}^m$, where δ_{ij} is Kronecker's delta symbol.

Let $\mathcal{L}_\alpha=Span\{e_i\}_{i\in\alpha}$ be a linear span of vectors $\{e_i\}_{i\in\alpha}$ and $\mathcal{S}^{m-1}[\mathcal{L}_\alpha]=\mathcal{S}^{m-1}\cap\mathcal{L}_\alpha$. One can easily check that $\mathcal{S}^{m-1}[\mathcal{L}_\alpha]\simeq\mathcal{S}^{|\alpha|-1}$. In what follows, we use $\mathcal{S}^{|\alpha|-1}$ instead of $\mathcal{S}^{m-1}[\mathcal{L}_\alpha]$.

It is clear that if $\mathbf{Q}\oplus\mathbf{L}\left(\mathcal{S}^{m-1}\right)\subset\mathcal{S}^{n-1}$ then one has that $\mathbf{Q}\oplus\mathbf{L}\left(\mathcal{S}^{|\alpha|-1}\right)\subset\mathcal{S}^{n-1}$ for any proper subset $\alpha\subset I_m$. Now, we are interesting into the reverse statement.

Problem 4.4. *Do these inclusions* $\mathbf{Q}\oplus\mathbf{L}\left(\mathcal{S}^{|\alpha_k|-1}\right)\subset\mathcal{S}^{n-1}$ *for some collections of proper subsets* $\alpha_k\subset I_m$ *imply* $\mathbf{Q}\oplus\mathbf{L}\left(\mathcal{S}^{m-1}\right)\subset\mathcal{S}^{n-1}$*?*

In general, the answer is negative. Let us give some examples.

Example 4.5. *Let* $\mathbf{Q}:\mathcal{S}^2\to\mathcal{S}^2$ *be the quadratic operator defined as follows*

$$\mathbf{Q}(x)=\left(x_1^2+x_2^2,\sqrt{2}(x_1x_3+x_2x_3),x_3^2\right).$$

Let $\mathcal{S}^2_{12}=\{x\in\mathcal{S}^2:x_3=0\}$, $\mathcal{S}^2_{13}=\{x\in\mathcal{S}^2:x_2=0\}$, *and* $\mathcal{S}^2_{23}=\{x\in\mathcal{S}^2:x_1=0\}$.
It is clear that $\mathbf{Q}(\mathcal{S}^2_{12})\subset\mathcal{S}^2$, $\mathbf{Q}(\mathcal{S}^2_{13})\subset\mathcal{S}^2$, *and* $\mathbf{Q}(\mathcal{S}^2_{23})\subset\mathcal{S}^2$. *However* $\mathbf{Q}(\mathcal{S}^2)\not\subset\mathcal{S}^2$. *In fact,*

$$\mathbf{Q}\left(\frac{1}{\sqrt{3}},\frac{1}{\sqrt{3}},\frac{1}{\sqrt{3}}\right)=\left(\frac{2}{3},\frac{2\sqrt{2}}{3},\frac{1}{3}\right)\notin\mathcal{S}^2.$$

We have a strong belief that the following conjecture should be true.

Conjecture 4.6. *Let* $\mathbf{Q}\oplus\mathbf{L}:\mathbb{R}^m\to\mathbb{R}^n$ *be a quadratic plus linear operator given by* (2.5). *Then one has that* $\mathbf{Q}\oplus\mathbf{L}(\mathcal{S}^{m-1})\subset\mathcal{S}^{n-1}$ *if and only if* $\mathbf{Q}\oplus\mathbf{L}\left(\mathcal{S}^{|\alpha_k|-1}\right)\subset\mathcal{S}^{n-1}$ *for any* $\alpha_k\subset I$ *with* $|\alpha_k|=3$.

Acknowledgments

The author would like to thank to prof. F. Mukhamedov for his attention on the paper. The author also acknowledges the IIUM Grant EDW B 13–019–0904 and the MOHE Grant ERGS13–025–0058 for the financial support. The author is grateful to the Junior Associate scheme of the Abdus Salam International Centre for Theoretical Physics, Trieste, Italy.

References

[1] T. Ando, M-D. Choi, Non-linear completely positive maps. Aspects of Positivity in Functional Analysis, 3–13, (1986).
[2] A.F. Beardon, Iteration of rational functions: Complex analytic dynamical system, Springer, (2000).
[3] A. Guterman, C.-K. Li, P. Semrl, Some general techniques on linear preserver problems. Lin. Alg. Appl. 315, 61–81 (2000).
[4] N. Johnston. Characterizing Operations Preserving Separability Measures via Linear Preserver Problems. Linear and Multilinear Algebra, 59(10):1171–1187, (2011).
[5] C. K. Li, S. Pierce, Linear preserver problems. The American Mathematical Monthly 108, 591–605 (2001).
[6] C. K. Li, N. K. Tsing, Linear preserver problems: A brief introduction and some special techniques. Linear Algebra and its Applications 162–164, 217–235 (1992).
[7] V. Moretti, Spectral Theory and Quantum Mechanics; With an Introduction to the Algebraic Formulation, Springer, (2013).
[8] F. Mukhamedov, A. Abduganiev, On pure quasi quantum quadratic operators of $M_2(\mathbb{C})$. Open Systems & Information Dynamics 20 (4) 1350018 (18 pages) (2013).

On Derivations Of Genetic Algebras

Farrukh Mukhamedov[1], **Izzat Qaralleh**[2]

[1,2] Department of Computational & Theoretical Sciences, Faculty of Science, International Islamic University Malaysia, Kuantan, Pahang, Malaysia

E-mail: [1]farrukh_m@iium.edu.my, [2]izzat_math@yahoo.com

Abstract. A genetic algebra is a (possibly non-associative) algebra used to model inheritance in genetics. In application of genetics this algebra often has a basis corresponding to genetically different gametes, and the structure constant of the algebra encode the probabilities of producing offspring of various types. In this paper, we find the connection between the genetic algebras and evolution algebras. Moreover, we prove the existence of nontrivial derivations of genetic algebras in dimension two.

1. Introduction

In mathematical genetics, genetic algebras are (possibly non-associative) used to model inheritance in genetic. In application of genetic this algebra often has a basis corresponding to genetically different gametes, and the structure constant of the algebra encode the probabilities of producing offspring of various types. There exist several classes of non-associative algebras (baric, evolution, Bernstein, train, stochastic, etc.), whose investigation has provided a number of significant contributions to theoretical population genetics. Such classes have been defined different times by several authors, and all algebras belonging to these classes are generally called genetic. In recent years many authors have tried to investigate the difficult problem of classification of these algebras. The most comprehensive references for the mathematical research done in this area are [1, 2, 3, 4].

In [1] an evolution algebra $\mathbb{A}$ associated to the free population is introduced and using this non-associative algebra many results are obtained in explicit form, e.g. the explicit description of stationary quadratic operators, and the explicit solutions of a nonlinear evolutionary equation in the absence of selection, as well as general theorems on convergence to equilibrium in the presence of selection.

In [3] a new type of evolution algebra is introduced. This algebra also describes some evolution laws of genetics and it is an algebra E over a field K with a countable natural basis $e_1, e_2, \ldots$ and multiplication given by $e_ie_i = \sum_j a_{ij}e_j$, $e_ie_j = 0$ if $i \neq j$. Therefore, e_ie_i is viewed as "self-reproduction". The derivation for evolution algebra E is defined as usual, i.e. a linear operator $d : E \to E$ is called a derivation if

$$d(u \circ v) = d(u) \cdot v + u \cdot d(v)$$

for all $u, v \in E$.

Note that for any algebra, the space $Der(E)$ of all derivations is a Lie algebra with the commutator multiplication.

In the theory of non-associative algebras, particularly, in genetic algebras, the Lie algebra of derivations of a given algebra is one of the important tools for studying its structure. There has been much work on the subject of derivations of genetic algebras ([5],[6],[7]). For evolution algebras the system of equations describing the derivations are given in [3].

In [8] it was showed that the multiplication is defined in terms of derivations, showing the significance of derivations in genetic algebras. Several genetic interpretations of derivation of genetic algebra are given in [9].

The paper is organized as follows. In section 2 we recall some definitions and theorems, which are needed in this paper. In section 3 we describe derivations of three dimensional genetic algebras. section 3 is devoted to show the connection between genetic and evolution algebras in dimension two. In section 4 we prove the existence of nontrivial derivations of genetic algebras in dimension two.

2. Preliminaries

Let $\mathfrak{g}$ be an algebra over the field $\mathbb{K}$. Assume that $\mathfrak{g}$ admits a basis $\{e_1, ..., e_n\}$ such that the multiplication constants $P_{ij,k}$ with respect to this basis, are given by

$$e_i \cdot e_j = \sum_{k=1}^{n} P_{ij,k} e_k.$$

We say that $\mathfrak{g}$ is *a genetic algebra* if the multiplication constants $P_{ij,k}$ satisfy

(i) $P_{ij,k} \geq 0$

(ii) $\sum_{k=1}^{n} P_{ij,k} = 1$.

In that case, the basis $\{e_1, ..., e_n\}$ is called *a natural basis.*
Let $(E, \cdot)$ be an algebra over a field $\mathbb{F}$. If it admits a basis $\{e_1, e_2, \dots\}$ such that

$$e_i \cdot e_j = 0, \quad for\ i \neq j, \qquad e_i \cdot e_i = \sum_k a_{i,k} e_k, \quad for\ any\ i,$$

then E is called an *evolution algebra.* By $\mathbf{A}$ we denote the structural matrix of E, i.e. $\mathbf{A} = (a_{ij})_{1 \leq i,j \leq n}$

Recall that the *derivation* on algebra A is a linear operator $d : A \to A$ such that

$$d(u \cdot v) = d(u) \cdot v + u \cdot d(v)$$

for all $u, v \in A$.

Theorem 2.1 *[10] Let $d : E \to E$ be a derivation of an evolution algebra E with non-singular* evolution matrix *in basis $\langle e_1, \dots, e_n \rangle$. Then the derivation d is zero.*

Theorem 2.2 *[10] Let E be an evolution algebra with structural matrix $\mathbf{A} = (a_{ij})_{1 \leq i,j \leq n}$ in the natural basis $e_1, \dots, e_n$, and rank$\mathbf{A} = n - 1$ such that $e_n e_n = b(e_1 e_1)$. Then the following assertions are hold true:*

(i) If $b = 0$, then the derivations d of E is either zero or it is in one of the following forms up to basis permutation:

$$\begin{pmatrix} 0 & \dots & 0 & d_{1n} \\ \vdots & \ddots & \vdots & \vdots \\ 0 & \dots & 0 & d_{n-1n} \\ 0 & \dots & 0 & 0 \end{pmatrix}, \tag{D_1}$$

where $\sum_{k=1}^{n-1} a_{ik}d_{kn} = 0,\ 1 \leq i \leq n-1;$

$$\begin{pmatrix} 0 & \dots & 0 & 0 & \dots & 0 & 0 \\ \vdots & \ddots & \vdots & \vdots & & \vdots & \vdots \\ 0 & \dots & 0 & 0 & \dots & 0 & 0 \\ 0 & \dots & 0 & \frac{d_{nn}}{2^{n-k-1}} & \dots & 0 & d_{k+1n} \\ \vdots & \ddots & \vdots & \vdots & \ddots & \vdots & \vdots \\ 0 & \dots & 0 & 0 & \dots & \frac{d_{nn}}{2} & d_{n-1n} \\ 0 & \dots & 0 & 0 & \dots & 0 & d_{nn} \end{pmatrix}, \tag{D_2}$$

where $d_{i+1n} = \frac{a_{in}}{a_{ii+1}}\left(\frac{1}{2^{n-i-1}} - 1\right)d_{nn},\ a_{ii+1} \neq 0,\ k+1 \leq i \leq n-2, 1 \leq k \leq n-1$ *and* $d_{k+1n} \in \mathbb{C}$.

(ii) If $b \neq 0$. *Then derivation* d *is either zero or it is in one of the following forms up to basis permutation:*

(i) (D_3), *where* $d_{11} = \dfrac{\delta}{2^{n-s}-1},\ 1 \leq s \leq n-1$ *and* $\delta^2 = -bd_{1n}^2$;

(ii) (D_4), *where* $d_{22} = \dfrac{1-2^{m-k}}{2^{k-1}}d_{11}, d_{11} = \dfrac{\delta}{2^{m-k+1}-1},\ 1 \leq k < m \leq n-1$ *and* $\delta^2 = -bd_{1n}^2$;

(iii) (D_5), *where* $d_{11} = \delta$ *and* $\delta^2 = -bd_{1n}^2$.

Here D_3, D_4 *and* D_5 *are respectively given By*

$$\begin{pmatrix} d_{11} & 0 & \dots & 0 & 0 & \dots & 0 & d_{1n} \\ 0 & 0 & \dots & 0 & 0 & \dots & 0 & 0 \\ \vdots & \vdots & \ddots & \vdots & \vdots & & \vdots & \vdots \\ 0 & 0 & \dots & 0 & 0 & \dots & 0 & 0 \\ 0 & 0 & \dots & 0 & 2d_{11} & \dots & 0 & 0 \\ \vdots & \vdots & & \vdots & \vdots & \ddots & \vdots & \vdots \\ 0 & 0 & \dots & 0 & 0 & \dots & 2^{n-s-1}d_{11} & 0 \\ -bd_{1n} & 0 & \dots & 0 & 0 & \dots & 0 & d_{11} \end{pmatrix} \tag{D_3}$$

$$\begin{pmatrix} d_{11} & 0 & \dots & 0 & 0 & \dots & 0 & 0 & \dots & 0 & d_{1n} \\ 0 & d_{22} & \dots & 0 & 0 & \dots & 0 & 0 & \dots & 0 & 0 \\ \vdots & \vdots & \ddots & \vdots & \vdots & & \vdots & \vdots & & \vdots & \vdots \\ 0 & 0 & \dots & 2^{k-1}d_{22} & 0 & \dots & 0 & 0 & \dots & 0 & 0 \\ 0 & 0 & \dots & 0 & 2d_{11} & \dots & 0 & 0 & \dots & 0 & 0 \\ 0 & 0 & \dots & \vdots & \vdots & \ddots & \vdots & \vdots & & \vdots & \vdots \\ 0 & 0 & \dots & 0 & 0 & \dots & 2^{m-k}d_{11} & 0 & \dots & 0 & 0 \\ 0 & 0 & \dots & 0 & 0 & \dots & 0 & 0 & \dots & 0 & 0 \\ \vdots & \vdots & \ddots & \vdots & \vdots & & \vdots & \vdots & \ddots & \vdots & \vdots \\ 0 & 0 & \dots & 0 & 0 & \dots & 0 & 0 & \dots & 0 & 0 \\ -bd_{1n} & 0 & \dots & 0 & 0 & \dots & 0 & 0 & \dots & 0 & d_{11} \end{pmatrix} \tag{D_4}$$

$$\begin{pmatrix} d_{11} & 0 & \dots & 0 & 0 & \dots & 0 & 0 & d_{1n} \\ 0 & 0 & \dots & 0 & 0 & \dots & 0 & 0 & 0 \\ \vdots & \vdots & \ddots & \vdots & \vdots & & \vdots & \vdots & \vdots \\ 0 & 0 & \dots & 0 & 0 & \dots & 0 & 0 & 0 \\ 0 & 0 & \dots & 0 & \frac{d_{11}}{2^{n-s-2}} & \dots & 0 & 0 & 0 \\ \vdots & \vdots & & \vdots & \vdots & \ddots & \vdots & \vdots & \vdots \\ 0 & 0 & \dots & 0 & 0 & \dots & \frac{d_{11}}{2} & 0 & 0 \\ 0 & 0 & \dots & 0 & 0 & \dots & 0 & d_{11} & 0 \\ -bd_{1n} & 0 & \dots & 0 & 0 & \dots & 0 & 0 & d_{11} \end{pmatrix} \qquad (D_5)$$

3. Derivations Of Three Dimensional Genetic Algebras

In this section, we are going to describe derivations of three dimensional genetic algebras. Let $\{e_1, e_2, e_3\}$ be a basis of three dimensional genetic algebra then the rule of multiplication is defined as follows:

$$(e_i \circ e_j) = \sum_{i,j=1}^{3} P_{ij,k} e_k, \quad k = \overline{1,3}, \qquad (1)$$

where $\{P_{ij,k}\}$ are coefficients of heredity, which satisfy the following conditions

$$P_{ij,k} \geq 0, \quad P_{ij,k} = P_{ji,k}, \quad \sum_{k=1}^{3} P_{ij,k} = 1, i, j, k \in \{1, 2, 3\}. \qquad (2)$$

Let us define a derivation of the genetic algebra. Then one can represent it as

$$d(e_i) = \sum_{j=1}^{3} d_{ij} e_j.$$

So, let us start to calculate

$$d(e_1 \circ e_1) = d(e_1)e_1 + e_1 d(e_1) = 2d(e_1)e_1.$$

Hence,

$$d(e_1 \circ e_1) = d(p_{11,1}e_1 + p_{11,2}e_2 + p_{11,3}e_2) = 2(d_{11}e_1 + d_{12}e_2 + d_{13}e_3)e_1,$$

then

$$p_{11,1}d(e_1) + p_{11,2}d(e_2) + p_{11,3}d(e_3) = 2(d_{11}(e_1.e_1) + d_{12}(e_2.e_1) + d_{13}(e_3.e_1).$$

Put the values of $d(e_i)$ and $(e_i \circ e_j)$ into above expression and compare the coefficients, we get

$$\begin{cases} p_{11,1}d_{11} + p_{11,2}d_{21} + p_{11,3}d_{31} = 2p_{11,1}d_{11} + 2p_{12,1}d_{12} + 2p_{13,1}d_{13} \\ p_{11,1}d_{12} + p_{11,2}d_{22} + p_{11,3}d_{32} = 2p_{11,2}d_{11} + 2p_{12,2}d_{12} + 2p_{13,2}d_{13} \\ p_{11,1}d_{13} + p_{11,2}d_{23} + p_{11,3}d_{33} = 2p_{11,3}d_{11} + 2p_{12,3}d_{12} + 2p_{13,3}d_{13} \end{cases} \qquad (3)$$

In the same manner one gets the following system, for $d(e_2 \circ e_2)$:

$$\begin{cases} p_{22,1}d_{11} + p_{22,2}d_{21} + p_{22,3}d_{31} = 2p_{12,1}d_{21} + 2p_{22,1}d_{22} + 2p_{23,1}d_{23} \\ p_{22,1}d_{12} + p_{22,2}d_{22} + p_{22,3}d_{32} = 2p_{12,2}d_{21} + 2p_{22,2}d_{22} + 2p_{23,2}d_{23} \\ p_{22,1}d_{13} + p_{22,2}d_{23} + p_{22,3}d_{33} = 2p_{12,3}d_{21} + 2p_{22,3}d_{22} + 2p_{23,3}d_{23} \end{cases} \qquad (4)$$

From $d(e_3 \circ e_3)$ one finds

$$\begin{cases} p_{33,1}d_{11} + p_{33,2}d_{21} + p_{33,3}d_{31} = 2p_{13,1}d_{31} + 2p_{23,1}d_{32} + 2p_{33,1}d_{33} \\ p_{33,1}d_{12} + p_{33,2}d_{22} + p_{33,3}d_{32} = 2p_{13,2}d_{31} + 2p_{23,2}d_{32} + 2p_{33,2}d_{33} \\ p_{33,1}d_{13} + p_{33,2}d_{23} + p_{33,3}d_{33} = 2p_{13,3}d_{31} + 2p_{23,3}d_{32} + 2p_{33,3}d_{33} \end{cases} \tag{5}$$

From $d(e_1 \circ e_2)$ we obtain

$$\begin{cases} p_{12,1}d_{11} + p_{12,2}d_{21} + p_{12,3}d_{31} &= p_{12,1}d_{11} + p_{22,1}d_{12} + p_{23,1}d_{13} + p_{11,1}d_{21} \\ &\quad + p_{12,1}d_{22} + p_{13,1}d_{23} \\ p_{12,1}d_{12} + p_{12,2}d_{22} + p_{12,3}d_{32} &= p_{12,2}d_{11} + p_{22,2}d_{12} + p_{23,2}d_{13} + p_{11,2}d_{21} \\ &\quad + p_{12,2}d_{22} + p_{23,1}d_{23} \\ p_{12,1}d_{13} + p_{12,2}d_{23} + p_{12,3}d_{33} &= p_{12,3}d_{11} + p_{22,3}d_{12} + p_{23,3}d_{13} + p_{11,3}d_{21} \\ &\quad + p_{12,3}d_{22} + p_{13,3}d_{23} \end{cases} \tag{6}$$

From $d(e_1 \circ e_3)$ one has

$$\begin{cases} p_{13,1}d_{11} + p_{13,2}d_{21} + p_{13,3}d_{31} &= p_{13,1}d_{11} + p_{23,1}d_{12} + p_{33,1}d_{13} + p_{11,1}d_{31} \\ &\quad + p_{12,1}d_{32} + p_{13,1}d_{33} \\ p_{13,1}d_{12} + p_{13,2}d_{22} + p_{13,3}d_{32} &= p_{13,2}d_{11} + p_{23,2}d_{12} + p_{33,2}d_{13} + p_{11,2}d_{31} \\ &\quad + p_{12,2}d_{32} + p_{13,1}d_{33} \\ p_{13,1}d_{13} + p_{13,2}d_{23} + p_{13,3}d_{33} &= p_{13,3}d_{11} + p_{23,3}d_{12} + p_{33,3}d_{13} + p_{11,3}d_{31} \\ &\quad + p_{12,3}d_{32} + p_{13,3}d_{33} \end{cases} \tag{7}$$

From $d(e_2 \circ e_3)$ one finds

$$\begin{cases} p_{23,1}d_{11} + p_{23,2}d_{21} + p_{23,3}d_{31} &= p_{12,1}d_{21} + p_{23,1}d_{22} + p_{33,1}d_{23} + p_{12,1}d_{31} \\ &\quad + p_{22,1}d_{32} + p_{23,1}d_{33} \\ p_{23,1}d_{12} + p_{23,2}d_{22} + p_{23,3}d_{32} &= p_{12,2}d_{21} + p_{23,2}d_{22} + p_{33,2}d_{23} + p_{12,2}d_{31} \\ &\quad + p_{22,2}d_{32} + p_{23,2}d_{33} \\ p_{23,1}d_{13} + p_{23,2}d_{23} + p_{23,3}d_{33} &= p_{12,3}d_{21} + p_{23,3}d_{12} + p_{33,3}d_{23} + p_{12,3}d_{31} \\ &\quad + p_{22,3}d_{32} + p_{23,3}d_{33} \end{cases} \tag{8}$$

Now, let us consider the following example

Example 3.1 *Let*

$$p_{ii,k} = \begin{cases} 0 & : i \neq k \\ 1 & : i = k \end{cases}$$

And when $i \neq j$ we have the following matrix

$$p_{ij,k} = \begin{pmatrix} a & a & 0 \\ 0 & 0 & 1 \\ 1-a & 1-a & 0 \end{pmatrix}, a \in [0,1]$$

Now, we are going to find that the derivations of genetic algebra corresponding to $\{p_{ij,k}\}$. Let us first substitute the values of $p_{ij,k}$ into (3),(4),(5),(6),(7), and (9), Hence, we obtain

$$\begin{cases} -d_{11} - 2ad_{12} - 2ad_{13} = 0 \\ d_{12} = 0 \\ d_{13} - (2(1-a))d_{12} - (2(1-a))d_{13} = 0 \\ d_{21} - 2ad_{21} = 0 \\ -d_{22} - 2d_{23} = 0 \\ d_{23} - (2(1-a))d_{21} = 0 \\ d_{31} - 2ad_{31} = 0 \\ -d_{32} = 0 \\ -d_{33} - (2(1-a))d_{31} = 0 \\ (1-a)d_{31} - d_{21} - ad_{22} - ad_{23} = 0 \\ ad_{12} + (1-a)d_{32} - d_{12} - d_{13} = 0 \\ ad_{13} + (1-a)d_{33} - (1-a)d_{11} - (1-a)d_{22} - (1-a)d_{23} = 0 \\ (1-a)d_{31} - d_{31} - ad_{32} - ad_{33} = 0 \\ ad_{12} + (1-a)d_{32} - d_{12} - ad_{33} = 0 \\ ad_{13} - (1-a)d_{11} - d_{13} - (1-a)d_{32} = 0 \\ d_{21} - ad_{21} - ad_{31} = 0 \\ -d_{32} - d_{33} = 0 \\ -(1-a)d_{21} - (1-a)d_{31} = 0 \end{cases} \tag{9}$$

Now, let us solve the above system. It is clear that $d_{12} = d_{32} = 0$. Therefore, from $-d_{32} - d_{33} = 0$ one gets $d_{33} = 0$. Now, from $(1-a)d_{31} - d_{31} - ad_{32} - ad_{33} = 0$ and $-d_{33} - (2(1-a))d_{31} = 0$. we drive $d_{31} = 0$. Also from $-(1-a)d_{21} - (1-a)d_{31} = 0$ and $d_{21} - 2ad_{21} = 0$ we obtain $d_{21} = 0$. Therefore, from $d_{23} - (2(1-a))d_{21} = 0$ one finds $d_{23} = 0$. Consequently, $d_{22} = 0$ since $-d_{22} - 2d_{23} = 0$. From $ad_{12} + (1-a)d_{32} - d_{12} - d_{13} = 0$ we obtain $d_{13} = 0$. Then $-d_{11} - 2ad_{12} - 2ad_{13} = 0$ implies that $d_{11} = 0$. So, the derivation is zero
It is interesting to know the existence of nontrivial derivations. So, we will show the existence of nontrivial derivations in section five.

4. Relation Between Genetic and Evolution algebras

From the definition of genetic algebra and evolution algebra one can ask the following question: Is there a transformation of genetic algebra to some evolution algebra?
In this section, we are going to answer to this question in dimension two. Let $(p_{ij,k})$ be a structure matrix of the genetic algebra in dimension two. Namely, if e_1, e_2 are the basis of genetic algebra i.e.

$$e_i \circ e_j = \sum_{k=1}^{2} p_{ij,k} e_k. \tag{10}$$

Now, we want to change the given basis e_1, e_2 to a new basis f_1, f_2 such that the algebra generated by f_1, f_2 becomes an evolution algebra.

Theorem 4.1 *Let $\mathfrak{g}$ be a genetic algebra generated by the basis e_1, e_2 with structure constant $(p_{ij,k})$. Then $\mathfrak{g}$ is an evolution algebra with respect to new basis f_1, f_2 if and only if one of the following conditions are satisfied:*

(i) $p_{21,1} = p_{12,1} \neq \frac{p_{11,1}+p_{22,1}}{2}$

(ii) $p_{11,1} = p_{12,1} = p_{22,1}$

Proof. Assume that the change of bases is given by $f_u = \sum_{j=1}^{2} t_{uj}e_j$. One can see that

$$\begin{aligned} f_1 \circ f_2 &= \left(\sum_{j=1}^{2} t_{1j}e_j\right) \circ \left(\sum_{l=1}^{2} t_{2l}e_l\right) \\ &= \sum_{j,\ l=1}^{2} t_{1j}t_{2l} \circ \sum_{k=1}^{2} p_{jl,k}e_k \\ &= \sum_{k=1}^{2}\left(\sum_{j,\ l=1}^{2} t_{1j}t_{2l}p_{jl,k}\right)e_k \end{aligned} \tag{11}$$

Due to the definition evolution algebra we conclude that an algebra generated by f_1, f_2 is evolution if and only if $f_1 \circ f_2 = 0$. So, from (11) we infer that $f_1 \circ f_2 = 0$ if and only if $\sum_{j,\ l=1}^{2} t_{1j}t_{2l}p_{jl,k} = 0$. Let us redefine the matrix (t_{il}) as follows $(t_{jl}) = \begin{pmatrix} x & y \\ u & v \end{pmatrix}$, and explicitly rewrite the expression $\sum_{j,\ l=1}^{2} t_{1j}t_{2l}p_{jl,k} = 0$. Hence, one finds

$$p_{11,1}xu + p_{12,1}xv + p_{21,1}yu + p_{22,1}yv = 0 \tag{12}$$

$$(1 - p_{11,1})xu + (1 - p_{12,1})xv + (1 - p_{21,1})yu + (1 - p_{22,1})yv = 0 \tag{13}$$

By adding (12) and (13) we obtain

$$xu + xv + yu + yv = 0.$$

So, $(u + v)(x + y) = 0$. This means $y = -x$ or $u + v = 0$. without loss of generosity we assume $y = -x$

Now, let $y = -x$ then the matrix (t_{jl}), after making a simple scale, takes the following form

$$(t_{jl}) = \begin{pmatrix} 1 & -1 \\ u & v \end{pmatrix}$$

And by back substituting into (12) we have

$$p_{11,1}u + p_{12,1}v - p_{21,1}u - p_{22,1}v = 0. \tag{14}$$

Now consider several cases

Case 1. Let $u = 0$, then $v \neq 0$, since $det(t_{jl}) \neq 0$. Therefore, from (14) one has $(p_{12,1} - p_{22,1})v = 0$ which yields $p_{12,1} = p_{22,1}$. So, the matrix (t_{jl}) has the following form

$$(t_{jl}) = \begin{pmatrix} 1 & -1 \\ 0 & 1 \end{pmatrix}$$

Case 2. Let $u \neq 0$ and $p_{12,1} = p_{22,1}$. Then from (14) we have $(p_{11,1} - p_{12,1})u = 0$. Therefore, $p_{11,1} = p_{12,1} = p_{12,1}$. So, the matrix (t_{jl}) takes the following form

$$(t_{jl}) = \begin{pmatrix} 1 & -1 \\ 1 & 0 \end{pmatrix}$$

Case 3. Let $u \neq 0$ and $p_{12,1} \neq p_{22,1}$. Then from (14) one finds

$$p_{11,1} + p_{12,1}\left(\frac{v}{u}\right) - p_{21,1} - p_{22,1}\left(\frac{v}{u}\right) = 0$$

By a simple algebra we get

$$\frac{v}{u} = \frac{p_{21,1} - p_{11,1}}{p_{12,1} - p_{22,1}}.$$

Therefore, $v = \frac{p_{21,1}-p_{11,1}}{p_{12,1}-p_{22,1}} \cdot u$ Then the matrix (t_{jl}), by making simple scale, takes the following form:

$$(t_{jl}) = \begin{pmatrix} 1 & -1 \\ u & \frac{p_{21,1}-p_{11,1}}{p_{12,1}-p_{22,1}}.u \end{pmatrix}$$

If $p_{11,1} + p_{22,1} \neq 2p_{12,1}$, then the matrix (t_{jl}) is non singular,i.e. $det(t_{ij}) \neq 0$, and it teaks the following form

$$(t_{jl}) = \begin{pmatrix} 1 & -1 \\ 1 & \frac{p_{21,1}-p_{11,1}}{p_{12,1}-p_{22,1}} \end{pmatrix}.$$

This completes the proof.

5. Derivations of genetic algebras in dimension two

Its natural to know the existence of nontrivial derivations of genetic algebras. In this section by means of the pervious section we are able to provide some conditions on structure constants of genetic algebra for that algebra exists a nontrivial derivation.

Theorem 5.1 *Let $\mathfrak{g}$ be a genetic algebra with basis $\{e_1, e_2\}$ of dimension two.*

(i) If $p_{21,1} = p_{12,1} \neq \frac{p_{11,1}+p_{22,1}}{2}$. Then any derivation is trivial,i.e. zero.

(ii) If $p_{11,1} = p_{12,1} = p_{22,1}$. Then there exists a nontrivial derivation.

Proof. Let $\mathfrak{g}$ be a genetic algebra. Now, we choose such a basis f_1, f_2 in $\mathfrak{g}$ such that $\mathfrak{g}$ becomes an evolution algebra. Due theorem 4.1, this occurs if $(i), (ii)$ are satisfied.

(i) Let $p_{12,1} = p_{21,1} \neq \frac{p_{11,1}+p_{22,1}}{2}$. Then we have the following transformation matrix

$$(t_{jl}) = \begin{pmatrix} 1 & -1 \\ 1 & \frac{p_{21,1}-p_{11,1}}{p_{12,1}-p_{22,1}} \end{pmatrix}$$

One can see that $f_1 = e_1 - e_2$, $f_2 = e_1 - \alpha e_2$ where $\alpha = \frac{p_{21,1}-p_{11,1}}{p_{12,1}-p_{22,1}}$. Then by a simple algebra we find

$$\begin{aligned} f_1^2 &= \left(\frac{\gamma\alpha}{1+\alpha} - \frac{\beta}{1+\alpha}\right) f_1 + \left(\frac{\gamma}{1+\alpha} + \frac{\beta}{1+\alpha}\right) f_2 \\ f_2^2 &= \left(\frac{x\alpha}{1+\alpha} - \frac{y}{1+\alpha}\right) f_1 + \left(\frac{x}{1+\alpha} + \frac{y}{1+\alpha}\right) f_2 \end{aligned}$$

Where $\gamma = p_{11,1} + p_{22,1} - 2p_{12,1}, \beta = p_{11,2} + p_{22,2} - 2p_{12,2}, x = p_{11,1} + 2\alpha p_{12,1} + \alpha^2 p_{22,1}$ and $y = p_{11,2} + 2\alpha p_{12,2} + \alpha^2 p_{22,2}$. So, we have the following structural matrix of evolution algebra

$$\mathbf{A} = \begin{pmatrix} \gamma\alpha - \beta & \gamma + \beta \\ x\alpha - y & x + y \end{pmatrix}$$

Suppose that $det(A) = 0$, then $det(A) = (\alpha + 1)(\gamma y - \beta x) = 0$. If $\alpha = -1$ then one can find $p_{12,1} = p_{21,1} = \frac{p_{11,1}+p_{22,1}}{2}$, which is a contradiction to our case. So, $\frac{\gamma}{\beta} = \frac{x}{y}$, i.e. $\frac{p_{11,1}+p_{22,1}-2p_{12,1}}{1-p_{11,1}+1-p_{22,1}-2+2p_{12,1}} = \frac{x}{y}$ therefore, $x = -y$ then

$$p_{11,1} + 2\alpha p_{12,1} + \alpha^2 p_{22,1} = -1 + p_{11,1} - 2\alpha(1 - p_{12,1}) - \alpha^2(1 - p_{22,1}),$$

this means that $(\alpha + 1)^2 = 0$, which is impossible. Hence, $det(A) \neq 0$. Then by theorem 2.1 any derivation is zero in this case.

(ii) Let $p_{11,1} = p_{12,1} = p_{22,1}$ then we have the following matrix

$$(t_{jl}) = \begin{pmatrix} 1 & -1 \\ 1 & 0 \end{pmatrix}$$

Therefore, one finds $f_1 = e_1 - e_2,\ f_2 = e_1$. Then by simple algebra we get

$$\begin{aligned} f_1^2 &= -\beta f_1 + (\gamma + \beta) f_2 \\ f_2^2 &= -p_{11,2} f_1 + f_2, \end{aligned}$$

where $\gamma = p_{11,1} + p_{22,1} - 2p_{12,1}, \beta = p_{11,2} + p_{22,2} - 2p_{12,2}$. So, we have the following structure matrix of evolution algebra:

$$\mathbf{A} = \begin{pmatrix} -\beta & \gamma + \beta \\ -p_{11,2} & 1 \end{pmatrix}$$

But $\gamma = p_{11,1} + p_{22,1} - 2p_{12,1} = 0 = -\beta$. Therefore, the $det(A) = 0$. Then according to Theorem 2.2, we have a nontrivial derivation. This completes the proof.

Corollary 5.2 *Let* $\mathfrak{g}$ *be a genetic algebra with basis* $\{e_1, e_2\}$. *Then the nontrivial derivations genetic algebra take the following form*

$$\overline{d} = \begin{pmatrix} 0 & d_{12} \\ -d_{12} & 0 \end{pmatrix}$$

Proof. By Theorem 2.2 we have the following nontrivial derivation evolution algebra

$$d = \begin{pmatrix} 0 & d_{12} \\ 0 & 0 \end{pmatrix}.$$

Since in the Theorem 5.1 we have $f_1 f_1 = 0$, then the nontrivial derivation evolution algebra

$$d = \begin{pmatrix} 0 & d_{12} \\ 0 & 0 \end{pmatrix}.$$

takes the following form

$$d_1 = \begin{pmatrix} 0 & 0 \\ d_{12} & 0 \end{pmatrix}.$$

and by using Theorem 5.1(ii) one has

$$\overline{d} = \begin{pmatrix} 1 & -1 \\ 1 & 0 \end{pmatrix} \begin{pmatrix} 0 & 0 \\ d_{12} & 0 \end{pmatrix} \begin{pmatrix} 0 & 1 \\ -1 & 1 \end{pmatrix} = \begin{pmatrix} 0 & d_{12} \\ -d_{12} & 0 \end{pmatrix}$$

This completes the proof.

Acknowledgments
The authors acknowledge the MOHE grant ERGS 13-024-0057 for the financial support.

References

[1] Lyubich Y I 1992 *Mathematical structures in population genetics* (Springer-Verlag, Berlin)
[2] Reed M L 1997 Bull. Amer. Math. Soc. (N.S.) **34(2)** 107–130.
[3] Tian J P 1921 *Evolution algebras and their applications*, Lecture Notes in Mathematics (Springer-Verlag, Berlin, 2008).
[4] Wörz-Busekros A 1980 *Algebras in genetics*, Lecture Notes in Biomathematics (Springer-Verlag, Berlin-New York) .
[5] Costa R 1982 Bol. Soc. Brasil. Mat. **13** 69–81 .
[6] Costa R 1983 Bol. Soc. Brasil. Mat. **14** 63–80.
[7] Gonshor H 1988 **16 (8)** 1525–1542 .
[8] Micali A, Revoy P 1986 Proc. Edinburgh Math. Soc. **29 (2)** 187–197.
[9] Holgate P 1987 Linear Algebra Appl **85** 75–79.
[10] Camacho L M, Gmez J R ,Omirov B A and Turdibaev R M 2013 Linear and Multilinear Algebra **61, no. 3** 309-322.

Canonical Groups for Quantization on the Two-Dimensional Sphere and One-Dimensional Complex Projective Space

Sumadi A H A[1] and Zainuddin H[1,2]

[1] Department of Physics, Faculty of Science, UPM, 43400, Serdang Selangor.
[2] Laboratory of Computational Science and Mathematical Physics, INSPEM, UPM, 43400, Serdang Selangor.

E-mail: ahmadhazazi@gmail.com

Abstract. Using Isham's group-theoretic quantization scheme, we construct the canonical groups of the systems on the two-dimensional sphere and one-dimensional complex projective space, which are homeomorphic. In the first case, we take SO(3) as the natural canonical Lie group of rotations of the two-sphere and find all the possible Hamiltonian vector fields, and followed by verifying the commutator and Poisson bracket algebra correspondences with the Lie algebra of the group. In the second case, the same technique is resumed to define the Lie group, in this case SU (2), of CP^1.We show that one can simply use a coordinate transformation from S^2 to CP^1 to obtain all the Hamiltonian vector fields of CP^1. We explicitly show that the Lie algebra structures of both canonical groups are locally homomorphic. On the other hand, globally their corresponding canonical groups are acting on different geometries, the latter of which is almost complex. Thus the canonical group for CP^1 is the double-covering group of SO(3), namely SU(2). The relevance of the proposed formalism is to understand the idea of CP^1 as a space of where the qubit lives which is known as a Bloch sphere.

1. Introduction

Group-theoretic quantization technique was proposed by Isham [1], [2] is a quantization procedure that is geometrical in nature but with emphasis on the group-theoretical aspects. The idea is to base the quantization primarily on a canonical group,G that describes the symmetries of the phase-space. The canonical group plays the role of a global analogue of the canonical commutation relations in canonical quantization. Due to this idea, an irreducible unitary representation of G, then shows the appropriate quantization of the system. This technique has been used widely by many authors such as [3], [4], [5] etc.

The aim of this paper is to adopt this technique to quantize the simple systems on S^2 and CP^1, which are both topologically homeormophic to each other. The cases considered are different from those which Isham and others have discussed in the sense that the phase spaces are not cotangent bundles of some configuration space but they are compact symplectic manifolds. In order to make this paper is more informative, a summary of results for both cases are discussed.

2. Discussions

2.1. *Quantization on the phase space $S=S^2$*

Instead of choosing a canonical group of the phase space, $S = S^2$, we first begin by identifying the set of (globally-defined) canonical observables analogous to the coordinates q^i and p^j in the conventional phase spaces. These observables produce the set of Hamiltonian vector fields on phase space whose integral curves are generated by the one-parameter subgroups of the canonical group. This procedure is then carried out again in the case of CP^1.We choose G= SO(3) as a candidate of a canonical group of S^2. Then, let us consider the phase space, $S = S^2$ is parametrized by the angles θ and $\emptyset$. This phase space is endowed with the natural symplectic structure

$$\omega = sin\theta \, d\theta \wedge d\emptyset \tag{1}$$

where $(\boldsymbol{\theta}, \emptyset)$ is the spherical coordinates. So, one could choose the following set of the globally well-defined functions on $S = S^2$ as

$$u = \boldsymbol{sin\,\theta cos\,\emptyset} \quad ; \tag{2}$$

$$v = \boldsymbol{sin\theta sin\emptyset} \quad ; \tag{3}$$

$$w = \boldsymbol{cos\theta} \tag{4}$$

to be the canonical observables. From this set of canonical observables one can obtained the set of corresponding Hamiltonian vector fields $\mathcal{E}_f$ by the relation $\mathcal{E}_f \lrcorner \boldsymbol{\omega} = df$, where f are the observables and $\mathcal{E}_f$ are their Hamiltonian vector fields. From this, they are:

$$\mathcal{E}_u = -\boldsymbol{sin\phi}\frac{\partial}{\partial\theta} - \boldsymbol{cot\theta cos\phi}\frac{\partial}{\partial\phi} \quad ; \tag{5}$$

$$\mathcal{E}_v = \boldsymbol{cos\phi}\frac{\partial}{\partial\theta} - \boldsymbol{cot\theta sin\phi}\frac{\partial}{\partial\phi} \quad ; \tag{6}$$

$$\mathcal{E}_w = \frac{\partial}{\partial\phi} \quad . \tag{7}$$

After obtaining the set of Hamiltonian vector fields, we could construct their commutator algebra which is as follows

$$[\mathcal{E}_u, \mathcal{E}_v] = -\mathcal{E}_w \quad ; \quad [\mathcal{E}_w, \mathcal{E}_u] = \mathcal{E}_v \quad ; \tag{8}$$

$$[\mathcal{E}_v, \mathcal{E}_w] = -\mathcal{E}_u \quad ; \quad [\mathcal{E}_v, \mathcal{E}_u] = \mathcal{E}_w \quad ;$$

$$[\mathcal{E}_u, \mathcal{E}_w] = -\mathcal{E}_v \quad ; \quad [\mathcal{E}_u, \mathcal{E}_u] = [\mathcal{E}_v, \mathcal{E}_v] = [\mathcal{E}_w, \mathcal{E}_w] = 0$$

$$[\mathcal{E}_w, \mathcal{E}_v] = \mathcal{E}_u$$

The Poisson bracket algebra of the observables is

$$\{\boldsymbol{u}, \boldsymbol{v}\} = \boldsymbol{w} \quad ; \quad \{\boldsymbol{w}, u\} = \boldsymbol{v} \quad ; \tag{9}$$

$$\{\boldsymbol{v}, \boldsymbol{w}\} = \boldsymbol{u} \quad ; \quad \{\boldsymbol{v}, \boldsymbol{u}\} = -\boldsymbol{w} \quad ;$$

$$\{\boldsymbol{u}, \boldsymbol{w}\} = -\boldsymbol{v} \quad ; \quad \{\boldsymbol{u}, u\} = \{\boldsymbol{v}, \boldsymbol{v}\} = \{\boldsymbol{w}, \boldsymbol{w}\} = 0;$$

$$\{\boldsymbol{w}, \boldsymbol{v}\} = -\boldsymbol{u} \quad .$$

The commutator algebra of the Hamiltonian vector fields is thus anti-homomorphic to the Poisson bracket algebra. Their forms suggest that they are further (anti-)homomorphic to the abstract algebra of the SO(3) and thus our choice of the canonical group.

An alternate route is to obtain the action of the group on the phase space, $S = S^2$ given by

$$\boldsymbol{\ell}(\mathrm{R}, \mathrm{s}) = \mathrm{Rs} \; ; \forall \, \mathrm{R} \in \boldsymbol{G} \, , \mathrm{s} \in S^2 \; . \tag{10}$$

We can generate the same Hamiltonian vector fields as before through their integral curves given by the group action. To conclude, SO(3) is considered to be the natural choice of the canonical group on S^2 and one can proceed to find its irreducible representation for further steps of quantization (this is not discussed here)

2.2 *Quantization on the phase space* $S = CP^1$

As we claimed that the homeomorphism of both phase spaces CP^1 and S^2 lead us to assume $\boldsymbol{G}$= SU(2) as the natural candidate of the canonical group of $S = CP^1$, due to the homomorphism of the Lie algebras of SO(3) and SU(2).

It is convenient to use the same technique as mentioned before to verify the canonical group of the phase space, $S = CP^1$. In order to simplify the calculations we shall work within the globally well-defined projective complex coordinates z and its conjugate z* on the sphere homeomorphic to CP^1 which is given by coordinate transformation,

$$Z = \boldsymbol{tan}\frac{\boldsymbol{\theta}}{2}\boldsymbol{e}^{\boldsymbol{i\phi}} \text{ and } Z^* = \mathbf{tan}\frac{\boldsymbol{\theta}}{2}\boldsymbol{e}^{-\boldsymbol{i\phi}} \; , \tag{11}$$

where $(\boldsymbol{\theta}, \emptyset)$ are the earlier coordinates of S^2. One can use the natural symplectic structure on the sphere together with (11) to form the complex symplectic structure on CP^1 given by

$$\boldsymbol{\omega} = \frac{\mathbf{2}\boldsymbol{i}\, \boldsymbol{dZ} \wedge \boldsymbol{dZ*}}{(\mathbf{1}+|\boldsymbol{Z}|^{\mathbf{2}})^{\mathbf{2}}} \; . \tag{12}$$

For simplicity the coordinate transformations could also be used to construct the canonical observables from (2) – (4). Hence we have

$$u' = \frac{Z+Z*}{1+|Z|^2} \quad ; \tag{13}$$

$$v' = \frac{-iZ+iZ*}{1+|Z|^2} \quad ; \tag{14}$$

$$w' = \frac{1-|Z|^2}{1+|Z|^2} \quad , \tag{15}$$

and these observables will further lead to the Hamiltonian vector fields as

$$\mathcal{E}_{u'} = -\frac{i}{2}(1 - Z^2)\frac{\partial}{\partial Z} + \frac{i}{2}(1 - Z^{*2})\frac{\partial}{\partial Z*} \quad ; \tag{16}$$

$$\mathcal{E}_{v'} = \frac{1}{2}(1 + Z^2)\frac{\partial}{\partial Z} + \frac{1}{2}(1 + Z^{*2})\frac{\partial}{\partial Z^*} \quad ; \tag{17}$$

$$\mathcal{E}_{w'} = iZ\frac{\partial}{\partial Z} - iZ^*\frac{\partial}{\partial Z^*} \quad . \tag{18}$$

From the above vector fields one can get the following commutator algebra as

$$[\mathcal{E}_{u'}, \mathcal{E}_{v'}] = -\mathcal{E}_{w'} \quad ; \quad [\mathcal{E}_{w'}, \mathcal{E}_{u'}] = \mathcal{E}_{v'} \quad ; \tag{19}$$

$$[\mathcal{E}_{v'}, \mathcal{E}_{w'}] = -\mathcal{E}_{u'} \quad ; \quad [\mathcal{E}_{v'}, \mathcal{E}_{u'}] = \mathcal{E}_{w'} \ ;$$

$$[\mathcal{E}_{u'}, \mathcal{E}_{w'}] = -\mathcal{E}_{v'} \quad ; \quad [\mathcal{E}_{u'}, \mathcal{E}_{u'}] = [\mathcal{E}_{v'}, \mathcal{E}_{v'}] = [\mathcal{E}_{w'}, \mathcal{E}_{w'}] = 0 \ ;$$

$$[\mathcal{E}_{w'}, \mathcal{E}_{v'}] = \mathcal{E}_{u'} \ ,$$

and the Poisson bracket algebra of the observables as

$$\{u', v'\} = w' \quad ; \quad \{w', u'\} = v' \quad ; \tag{20}$$

$$\{v', w'\} = u' \quad ; \quad \{v', 'u\} = -w' \ ;$$

$$\{u', w'\} = -v' \ ; \quad \{u', u'\} = \{v', v'\} = \{w', w'\} = 0 \ ;$$

$$\{w', v'\} = -u' \ .$$

The algebraic relations are the same as in (8) and (9). Note that in (13)-(15) and (16)-(18), there is a hidden discrete symmetry under the transformations $Z \to Z^*$, $i \to -i$. The discrete symmetry here shows that there is a hidden discrete symmetry on the phase-space that keeps the symplectic structure of CP^1 invariant. This may have implications in the global structure of the desired canonical group. On the other hand, we can construct the canonical group by exponentiating the vector fields and act upon the space of functions (square-integrable function). This gives a particular representation of the group. It is expected that the canonical group goes to the double cover of SO(3) i.e. SU(2). Presently details of this necessary lifting are being worked out.

3. Conclusions

In this study, we have shown that the group-theoretic quantization problems of the two-sphere and the one-dimensional complex projective space are related to each other through their algebraic structures. However, at the global level, they give different canonical groups which reflect the different underlying geometry.

Note that the present study only deals with the simplest case of the complex projective space and it is connected to the well-known example of (geometric) quantization on the two-sphere. This motivates us to consider the framework of Isham's group-theoretic quantization for quantizing complex projective spaces in general. In [12], the complex projective spaces are taken to be the classical phase spaces corresponding to the quantum states in quantum mechanics. It is thus then possible to use this framework to describe multiple qubit or qudit states as arising (geometrically) from a quantization problem and in particular aspects of entanglement.

References

[1] C.J. Isham 1984 *Relativity, Groups and Topology II*, eds. B.S. DeWittt and R.Stora, (North-Holland, Amsterdam) p. 1061-1290
[2] C. J. Isham and A. C. Kakas 1984 A group theoretical approach to the canonical quantisation of gravity.I Construction of the canonical group. Class. Quantum Grav. **1** 621-632.
[3] H. Zainuddin 1990 *Group-Theoretic Quantisation and Central Extensions, PhD Thesis* (University of Durham,England)
[4] A. Bouketir 2001 *Group-Theoretic Quantization on Spheres and Quantum Hall Effect, PhD Thesis* (UPM, Malaysia)
[5] C.L. Benavides and A.F. Reyes-Lega 2008 Canonical Group Quantization, Rotation Generators and Quantum Indistinguishability, (*Preprints math-ph/0806.2449v1*).
[6] C.J Isham 1999 *Modern Differential Geometry for Physicists*.(Singapore: World Scientific).
[7] B Y Hou and B Y Hou 1997 *Differential Geometry for Physicists*. (Singapore: World Scientific).
[8] A. Corichi and M.P. Ryan Jr 1997 Quantization of non-standard Hamiltonian systems *J .Phys. A: Math. Gen.* **30** 3553-3572
[9] B.C. Hall 2002 *Lie Groups, Lie Algebras and Representations An Elementary Introduction*. (Berlin: Springer)
[10] M. Nakahara 2003 *Geometry, Topology and Physics*, Second Ed. (IOP Publishing Bristol and Philadelphia)
[11] R. Abraham and J. E. Marsden 1985 *Foundation of Mechanics,* (Benjamin-Cummings, Reading, Massachusetts).
[12] J.M. Isidro 2004 Quantum states from tangent vectors Mod. Phys. Lett. A **19** 2339-2352

Diassociative algebras and their derivations

I M Rikhsiboev[1], **I S Rakhimov**[2] **and W Basri**[3]

[1]Universiti Kuala Lumpur, Malaysian Institute of Industrial Technology, 81750 Bandar Seri Alam, Johor Bahru, Malaysia

[2,3] Institute for Mathematical Research and Department of Mathematics, Faculty of Science, Universiti Putra Malaysia, 43400 UPM Serdang, Selangor Darul Ehsan, Malaysia

E-mail: [1]`ikromr@gmail.com`, [2]`risamiddin@gmail.com`, [3]`witriany@upm.edu.my`

Abstract. The paper concerns the derivations of diassociative algebras. We introduce one important class of diassociative algebras, give simple properties of the right and left multiplication operators in diassociative algebras. Then we describe the derivations of complex diassociative algebras in dimension two and three.

1. Introduction

Leibniz algebras and associative dialgebras (dialgebra) first arose in K-theory and rapidly became an object of great interest of many researchers. In 1993, J.Loday introduced the notion of Leibniz algebra [7] which is generalization of Lie algebra, where the skew-symmetricity of the brackets is dropped and the Jacobi identity is replaced by the so-called Leibniz identity. Loday also showed that the link between Lie algebras and associative algebras can be extended to an analogous relationship between Leibniz algebras and the so-called dialgebras (see [8]). Note that the dialgebra is generalization of associative algebra equipped with two products. In particular, it is easy can be shown that if on a vector space V two products $\dashv$ and $\vdash$ are given then the bracket $[\cdot,\cdot]$ by $[x,y] = x \dashv y - x \vdash y$ defines a Leibniz algebra structure on V. Conversely, the enveloping algebra of a Leibniz algebra has the structure of a dialgebra.

In the present paper we deal with the problem of description of derivations of diassociative algebras. The concept of derivation in this case can be easily imitated from that of finite-dimensional algebras. The algebra of derivations plays important role in the classification problems and in different applications of algebras. It is easy to show that the set of all derivations of a diassociative algebra form a Lie algebra with respect to the commutator bracket. In the paper we make use of classification results of two and three-dimensional complex diassociative algebras from [13].

Definition 1.1 *An associative dialgebra (or diassociative algebra) D over a field K is a vector space V over the K equipped with two bilinear associative binary operations denoted by $\dashv$ and $\vdash$, respectively, satisfying the following axioms:*

$$(x \dashv y) \dashv z = x \dashv (y \vdash z), \quad (x \vdash y) \dashv z = x \vdash (y \dashv z), \quad (x \dashv y) \vdash z = x \vdash (y \vdash z) \tag{1}$$

$\forall\ x, y, z \in V$, that is the triple $D = (V, \dashv, \vdash)$ with the axioms above is said to be a diassociative algebra.

Let us consider a few examples of diassociative algebras appeared in the literature.

Example 1.1 *Let $K[x,y]$ be the polynomial algebra over a field K of characteristic 0. If we define two multiplications on $K[x,y]$ as follows*

$$f(x,y) \dashv g(x,y) = f(x,y)g(y,y) \quad \text{and} \quad f(x,y) \vdash g(x,y) = f(x,x)g(x,y)$$

then $(K[x,y], \dashv, \vdash)$ is a diassociative algebra.

Example 1.2 *Let $(D, \dashv, \vdash)$ be a diassociative algebra. Consider the module of $n \times n$ - matrices $M_n(D) = M_n(K) \otimes D$ with products $(\alpha \dashv \beta)_{ij} = \Sigma_k \alpha_{ik} \dashv \beta_{kj}$ and $(\alpha \vdash \beta)_{ij} = \Sigma_k \alpha_{ik} \vdash \beta_{kj}$. Then $(M_n(D), \dashv, \vdash)$ is a diassociative algebra. Moreover, if D_1 and D_2 are diassociative algebras over a field K then their tensor product $D_1 \bigotimes_K D_2$ is provided by a dialgebra structure defined as follows:*

$$(a \otimes a') \star (b \otimes b') = (a \star b) \otimes (a' \star b') \quad \text{for} \quad \star = \vdash \text{ and } \dashv.$$

In fact, a diassociative algebra structure on an n-dimensional vector space V with a basis $\{e_1, e_2, ..., e_n\}$ can be given by defining the products of the basis vectors $\{e_1, e_2, ..., e_n\}$.

Example 1.3 *The products*

$$e_1 \vdash e_1 = e_1, \; e_1 \vdash e_2 = e_2, \; e_1 \dashv e_1 = e_1, \; e_2 \dashv e_1 = e_2$$

on two-dimensional and the products

$$e_1 \dashv e_2 = e_1, \; e_2 \dashv e_2 = e_2, \; e_3 \dashv e_3 = e_3, \; e_2 \vdash e_1 = e_1, \; e_2 \vdash e_2 = e_2, \; e_3 \vdash e_3 = e_3$$

on three-dimensional vector spaces define diassociative algebra structures, respectively.

A bar unit in D is an element $e \in D$ such that

$$x \dashv e = x = e \vdash x, \quad \text{for all } x \in D.$$

It is observed that the bar unit is not unique. The set of all bar units of a diassociative algebra is called a halo.

In Example 1.1 any element of the form $1 + (y - x)g(x,y)$ for $g(x,y) \in K[x,y]$ is a bar unit, therefore the halo of the diassociative algebra $K[x,y]$ is the subset $\{1 + (y-x)g(x,y) | g(x,y) \in F[x,y]\}$, meanwhile the identity matrix $I = \begin{pmatrix} 1 & 0 & ... & 0 \\ 0 & 1 & ... & 0 \\ ... & ... & ... & \\ 0 & 0 & ... & 1 \end{pmatrix}$ is a bar unit of $M_n(D) = M_n(K) \otimes D$ in Example 1.2 and e_1 is a bar unit in part one of Example 1.3.

Note also that there is no point unit (that is an element $e \in D$ such that $x \vdash e = e \dashv x = x$ for all $x \in D$) in a diassociative algebra, except for the case when the two products coincide, i.e., D is an associative algebra.

Definition 1.2 *A homomorphism of diassociative algebras D_1 and D_2 (provided both are given over the same field K) is a K-linear map $f : D_1 \to D_2$ such that $f(x \vdash y) = f(x) \vdash f(y)$ and $f(x \dashv y) = f(x) \dashv f(y)$ for all $x, y \in D_1$.*

As usual a bijective homomorphism is called isomorphism.

Definition 1.3 *A subspace D_0 of a diassociative algebra D is said to be subalgebra if $x \vdash y$ and $x \dashv y$ are in D_0 whenever $x, y \in D_0$.*

Journal of Physics: Conference Series **553** (2014) 012006 doi:10.1088/1742-6596/553/1/012006

Definition 1.4 *A two-sided ideal of a diassociative algebra D is a subspace I such that $x \star y$, $y \star x$ are in I for all $x \in D$, $y \in I$ with $\star = \vdash$ and $\dashv$. Note that I is called the right and left ideal if $y \vdash x$, $y \dashv x$ are in I, and $x \vdash y$, $x \dashv y$ are in I, respectively, for all $x \in D$, $y \in I$.*

Example 1.4 *Obviously, $I = \{0\}$ and D are two-sided ideals. As well as the kernel $Ker\varphi = \{x \in D_1 | \ \varphi(x) = 0\}$ of a homomorphism $\varphi : D_1 \longrightarrow D_2$ from diassociative algebra D_1 to D_2 is two-sided ideal in D_1 whereas the image $Im\varphi = \{y \in D_2 | \exists x \in D_1 : \ \varphi(x) = y\}$ is just a subalgebra of D_2.*

Let D be a Diassociative algebra and M, N be subsets of D. We define

$$M \Diamond N := M \vdash N + M \dashv N,$$

where

$$M \vdash N = Span_{\mathbb{C}}\{a \vdash b | \ a \in M, b \in N\}$$

and

$$M \dashv N = Span_{\mathbb{C}}\{a \dashv b | \ a \in M, b \in N\}.$$

It is obvious that if M is left (N is right) ideal in D so is $M \Diamond N$, respectively. Therefore, if both M and N are two-sided ideals so is $M \Diamond N$.

Let us consider the following series of two-sided ideals:

$$D^1 = D, \ D^{k+1} = D^1 \Diamond D^k + D^2 \Diamond D^{k-1} + ... + D^k \Diamond D^1 \qquad (2)$$

Definition 1.5 *A Diassociative algebra D is said to be nilpotent if there exists $s \in \mathbb{N}$ such that $D^s = 0$.*

Example 1.5 *Two dimensional algebra with multiplication table $e_1 \vdash e_1 = e_2$, $e_1 \dashv e_1 = \alpha e_2$, $\alpha \in \mathbb{C}$ on a basis $\{e_1, e_2\}$ is a nilpotent diassociative algebra.*

Note that the diassociative algebra in Example 1.2 and in Example 1.3 are not nilpotent.

An ideal I of diassociative algebra D is said to be nilpotent if it is nilpotent as a subalgebra of D.

It is observed that the sum $I_1 + I_2 = \{z \in D | z = x_1 + x_2, \ x_1 \in I_1$ and $x_2 \in I_2\}$ of two nilpotent ideals I_1, I_2 of D is nilpotent. Therefore there exists unique maximal nilpotent ideal of D called nilradical. The nilradical plays an important role in the classification problem of algebras.

Definition 1.6 *A derivation of diassociative algebra D is a linear transformation $d : D \to D$ satisfying*

$$d(x \dashv y) = d(x) \dashv y + x \dashv d(y) \ \text{and} \ d(x \vdash y) = d(x) \vdash y + x \vdash d(y)$$

for all $x, y \in D$.

The set of all derivations of a diassociative algebra D we denote by $Der(D)$. It is a Lie algebra with respect to the bracket $[d_1, d_2] = d_1 \circ d_2 - d_2 \circ d_1$.

Definition 1.7 *A dialgebra D is called characteristically nilpotent if elements of $Der(D)$ are nilpotent with respect to the composition.*

The study of characteristically nilpotent algebras is important in connection with the observations made by Jacobson in [3] and further developments of this concept for different types of algebraic structures we refer to [1], [2], [4], [5], [6], [9], [10], [11] and [12].

2. Results

Since a diassociative algebra possess two binary operations there are two right R_x, r_x and two left L_x, l_x multiplication operators defined as follows

$$R_x(y) := y \dashv x, \quad r_x(y) := y \vdash x,$$
$$L_x(y) := x \dashv y, \quad l_x(y) := x \vdash y.$$

Lemma 2.1 *The sets $R(D) = \{R_x | x \in D\}$, $L(D) = \{L_x | x \in D\}$, $r(D) = \{r_x | x \in D\}$, $l(D) = \{l_x | x \in D\}$ are closed with respect to the composition.*

Proof. The proof can be easily derived from the following identities

$$R_{x \dashv y} = R_y R_x, \qquad L_{x \dashv y} = L_y L_x,$$
$$r_{x \vdash y} = r_y r_x, \qquad l_{x \vdash y} = l_y l_x.$$

The proof of the next lemma also can be easily obtained by simple computations.

Lemma 2.2 *For the right and left multiplication operators of diassociative algebras the following identities hold true:*

$$R_x R_y = R_{r_x(y)}, \qquad R_x r_y = r_{R_x(y)}, \qquad r_x R_y = r_x r_y,$$
$$L_x L_y = L_x l_y, \qquad l_x L_y = L_{l_x(y)}, \qquad l_x l_y = l_{L_x(y)}.$$

Note that the following combinations of the right and left multiplication operators are also derivations of the diassociative algebra D:

$$L_x R_y + L_x L_y \text{ and } l_x r_y + l_x l_y.$$

Definition 2.1 *The subsets $Ann_R(D)$ and $Ann_L(D)$ defined by*

$$Ann_R(D) = \{x \in D | D \dashv x = 0, D \vdash x = 0\}$$

and

$$Ann_L(D) = \{x \in D | x \dashv D = 0, x \vdash D = 0\}$$

of a diassociative algebra D are called the right and the left annihilators of D, respectively.

Lemma 2.3 *The sets $Ann_R(D)$ and $Ann_L(D)$ are two-sided ideals of D.*

Let us consider diassociative algebra $(D, \vdash, \dashv)$ and linear transformations $ad_z(x) = x \dashv z - z \vdash x$. It is not difficult to verify that ad_z is a derivation of D. This type derivations are called inner derivations of diassociative algebra D. The set of all inner derivations we denote by $J(D)$. Then one has

Lemma 2.4 *The subset $J(D)$ is an ideal of the Lie algebra $Der(D)$.*

Proof. Indeed, $ad_{z_1} - ad_{z_2} = ad_{z_1 - z_2}$ and $[d, ad_z] = ad_{d(z)}$, for any $d \in Der(D)$.

Let D be an n-dimensional complex diassociative algebra and $\{e_1, e_2, ..., e_n\}$ be its basis. The components of $e_i \vdash e_j$ and $e_k \dashv e_s$, where $i, j, k, s = 1, 2, ..., n$ on the basis $\{e_1, e_2, ..., e_n\}$ are called the structure constants of D on $\{e_1, e_2, ..., e_n\}$, i.e., if

$$e_i \dashv e_j = \sum_{k=1}^{n} \gamma_{ij}^k e_k, \quad e_i \vdash e_j = \sum_{k=1}^{n} \delta_{ij}^k e_k$$

then the set

$$\{\gamma_{ij}^k, \delta_{st}^q \in K, 1 \leq i, j, k, s, t, q \leq n\}$$

is called the set of structure constants of D. This means that each point $\{\gamma_{ij}^k, \delta_{st}^q\}$ of the affine space K^{2n^3} defines an algebra structure on underlying vector space, however, for this structure to be a diassociative structure the scalars $\{\gamma_{ij}^k, \delta_{st}^q\}$ must satisfy conditions according to the axioms (1) of the diassociative algebra.

Further all the algebras considered are supposed to be over the field of complex numbers $\mathbb{C}$.

Let us now make a discussion on derivations of the diassociative algebras. A derivation d of D we represent in matrix form $d = (d_{ij})_{i,j=1,2,\dots,n}$ on the basis $\{e_1, e_2, \dots, e_n\}$. If the structure constants $\{\gamma_{ij}^k, \delta_{st}^q\}$ are given then we form a system of equations with respect to d_{ij} and solving this system we get the descriptions of the derivations.

This system has the following form:

$$\sum_{k=1}^{n} \gamma_{ij}^k d_{kt} = \sum_{k=1}^{n} (d_{ki}\gamma_{kj}^t + d_{kj}\gamma_{ik}^t), \quad \sum_{k=1}^{n} \delta_{ij}^k d_{kt} = \sum_{k=1}^{n} (d_{ki}\delta_{kj}^t + d_{kj}\delta_{ik}^t), \tag{2}$$

for $1 \leq i, j, t \leq n$.

Let us apply this approach to find the derivations of complex diassociative algebras in dimension two and three. We make use of classification results from [13].

In two dimensional case the system (2) has the following form:

$$\begin{aligned}
d_{12}\gamma_{11}^2 &= d_{21}\gamma_{21}^1 + d_{11}\gamma_{11}^1 + d_{21}\gamma_{12}^1,\\
d_{21}\gamma_{11}^1 + d_{22}\gamma_{11}^2 &= 2d_{11}\gamma_{11}^2 + d_{21}\gamma_{21}^2 + d_{21}\gamma_{12}^2,\\
d_{12}\gamma_{12}^2 &= d_{21}\gamma_{22}^1 + d_{12}\gamma_{11}^1 + d_{22}\gamma_{12}^1,\\
d_{21}\gamma_{12}^1 &= d_{11}\gamma_{12}^2 + d_{21}\gamma_{22}^2 + d_{12}\gamma_{11}^2,\\
d_{12}\gamma_{21}^2 &= d_{12}\gamma_{11}^1 + d_{22}\gamma_{21}^1 + d_{21}\gamma_{22}^1,\\
d_{21}\gamma_{21}^1 &= d_{12}\gamma_{11}^2 + d_{11}\gamma_{21}^2 + d_{21}\gamma_{22}^2,\\
d_{11}\gamma_{22}^1 + d_{12}\gamma_{22}^2 &= d_{12}\gamma_{12}^1 + 2d_{22}\gamma_{22}^1 + d_{12}\gamma_{21}^1,\\
d_{21}\gamma_{22}^1 &= d_{12}\gamma_{12}^2 + d_{22}\gamma_{22}^2 + d_{12}\gamma_{21}^2,
\end{aligned}$$

$$\begin{aligned}
d_{12}\delta_{11}^2 &= d_{21}\delta_{21}^1 + d_{11}\delta_{11}^1 + d_{21}\delta_{12}^1,\\
d_{21}\delta_{11}^1 + d_{22}\delta_{11}^2 &= 2d_{11}\delta_{11}^2 + d_{21}\delta_{21}^2 + d_{21}\delta_{12}^2,\\
d_{12}\delta_{12}^2 &= d_{21}\delta_{22}^1 + d_{12}\delta_{11}^1 + d_{22}\delta_{12}^1,\\
d_{21}\delta_{12}^1 &= d_{11}\delta_{12}^2 + d_{21}\delta_{22}^2 + d_{12}\delta_{11}^2,\\
d_{12}\delta_{21}^2 &= d_{12}\delta_{11}^1 + d_{22}\delta_{21}^1 + d_{21}\delta_{22}^1,\\
d_{21}\delta_{21}^1 &= d_{12}\delta_{11}^2 + d_{11}\delta_{21}^2 + d_{21}\delta_{22}^2,\\
d_{11}\delta_{22}^1 + d_{12}\delta_{22}^2 &= d_{12}\delta_{12}^1 + 2d_{22}\delta_{22}^1 + d_{12}\delta_{21}^1,\\
d_{21}\delta_{22}^1 &= d_{12}\delta_{12}^2 + d_{22}\delta_{22}^2 + d_{12}\delta_{21}^2.
\end{aligned}$$

The possible values of γ_{ij}^k and δ_{st}^q we take from the classification result of [13] mentioned above.

Theorem 2.1 *Any two-dimensional complex diassociative algebra is included in the following isomorphism classes*

$Dias_2^1 : e_1 \vdash e_1 = e_1,\ e_1 \dashv e_1 = e_1,\ e_2 \dashv e_1 = e_2;$
$Dias_2^2 : e_1 \vdash e_1 = e_1,\ e_1 \vdash e_2 = e_2,\ e_1 \vdash e_1 = e_1;$
$Dias_2^3 : e_1 \vdash e_1 = e_2,\ e_1 \dashv e_1 = \alpha e_2,\ \alpha \in \mathbb{C};$
$Dias_2^4 : e_1 \vdash e_1 = e_1,\ e_1 \vdash e_2 = e_2,\ e_1 \dashv e_1 = e_1,\ e_2 \dashv e_1 = e_2.$

Let us describe the derivations of $Dias_2^1$. Due to Theorem 2.1 we have $\gamma_{11}^1 = 1$, $\delta_{11}^1 = 1$, $\delta_{21}^2 = 1$ and other $\{\gamma_{ij}^k\ \delta_{st}^q\}$ are zero. Substituting and solving the system above we get the derivations of $Dias_2^1$ as follows

$$d = \begin{pmatrix} 0 & 0 \\ 0 & d_{22} \end{pmatrix}$$

The other cases can be easily found by the same way. As a result one has

Lemma 2.5 *The derivations of two dimensional complex diassociative algebras are given as follows*

Table 1. Derivations of two-dimensional diassociative algebras

Isomorphism Classes	Derivations	Dim. of the derivation algebras
$Dias_2^1$	$\begin{pmatrix} 0 & 0 \\ 0 & d_{22} \end{pmatrix}$	1
$Dias_2^2$	$\begin{pmatrix} 0 & 0 \\ 0 & d_{22} \end{pmatrix}$	1
$Dias_2^3(\alpha),\ \alpha \in \mathbb{C}$	$\begin{pmatrix} d_{11} & 0 \\ d_{21} & 2d_{11} \end{pmatrix}$	2
$Dias_2^4$	$\begin{pmatrix} 0 & 0 \\ d_{21} & d_{22} \end{pmatrix}$	2

Due to the result of [13] the isomorphism classes of three-dimensional diassociative algebras are given as follows.

Journal of Physics: Conference Series **553** (2014) 012006 doi:10.1088/1742-6596/553/1/012006

Theorem 2.2 *Any three-dimensional complex diassociative algebra is included in the following isomorphism classes of three-dimensional complex diassociative algebras*

$Dias_3^1 : e_1 \dashv e_2 = e_1, e_2 \dashv e_2 = e_2, e_3 \dashv e_3 = e_3, e_2 \vdash e_2 = e_2, e_3 \vdash e_3 = e_3;$

$Dias_3^2 : e_1 \dashv e_2 = e_1, e_2 \dashv e_2 = e_2, e_3 \dashv e_3 = e_3, e_2 \vdash e_1 = e_1, e_2 \vdash e_2 = e_2, e_3 \vdash e_3 = e_3;$

$Dias_3^3 : e_1 \dashv e_2 = e_1, e_2 \dashv e_2 = e_2, e_3 \dashv e_3 = e_3, e_2 \vdash e_2 = e_2, e_3 \vdash e_1 = e_1;$

$Dias_3^4 : e_1 \dashv e_3 = e_2, e_2 \dashv e_3 = e_2, e_3 \dashv e_3 = e_3, e_3 \vdash e_3 = e_3;$

$Dias_3^5 : e_1 \dashv e_3 = e_2, e_2 \dashv e_3 = e_2, e_3 \dashv e_3 = e_3, e_3 \vdash e_1 = e_1 - e_2, e_3 \vdash e_3 = e_3;$

$Dias_3^6 : e_1 \dashv e_3 = e_2, e_2 \dashv e_3 = e_2, e_3 \dashv e_3 = e_3, e_3 \vdash e_1 = e_1, e_3 \vdash e_2 = e_2, e_3 \vdash e_3 = e_3;$

$Dias_3^7 : e_1 \dashv e_3 = e_2, e_2 \dashv e_3 = e_2, e_3 \dashv e_3 = e_3, e_3 \vdash e_1 = e_2, e_3 \vdash e_2 = e_2, e_3 \vdash e_3 = e_3;$

$Dias_3^8 : e_1 \dashv e_3 = e_2, e_2 \dashv e_3 = e_2, e_3 \dashv e_3 = e_3, e_1 \vdash e_3 = e_2, e_2 \vdash e_3 = e_2,$
$e_3 \vdash e_1 = e_1 - e_2, e_3 \vdash e_3 = e_3;$

$Dias_3^9 : e_3 \dashv e_1 = e_2, e_3 \dashv e_2 = e_2, e_3 \dashv e_3 = e_3, e_3 \vdash e_1 = e_1, e_3 \vdash e_2 = e_2, e_3 \vdash e_3 = e_3;$

$Dias_3^{10} : e_3 \dashv e_1 = e_1, e_2 \dashv e_3 = e_2, e_3 \dashv e_3 = e_3, e_3 \vdash e_1 = e_1, e_3 \vdash e_3 = e_3;$

$Dias_3^{11} : e_3 \dashv e_1 = e_1, e_2 \dashv e_3 = e_2, e_3 \dashv e_3 = e_3, e_3 \vdash e_1 = e_1, e_3 \vdash e_2 = e_2, e_3 \vdash e_3 = e_3;$

$Dias_3^{12} : e_1 \dashv e_3 = e_1, e_2 \dashv e_3 = e_2, e_3 \dashv e_1 = e_1, e_3 \dashv e_3 = e_3, e_1 \vdash e_3 = e_1, e_3 \vdash e_1 = e_1,$
$e_3 \vdash e_3 = e_3;$

$Dias_3^{13} : e_1 \dashv e_3 = e_1, e_2 \dashv e_3 = e_2, e_3 \dashv e_1 = e_1, e_3 \dashv e_3 = e_3, e_1 \vdash e_3 = e_1, e_3 \vdash e_1 = e_1,$
$e_3 \vdash e_2 = e_2, e_3 \vdash e_3 = e_3;$

$Dias_3^{14} : e_1 \dashv e_3 = e_1, e_2 \dashv e_3 = e_2, e_3 \dashv e_1 = e_1, e_3 \dashv e_3 = e_3, e_1 \vdash e_3 = e_1 + e_2,$
$e_3 \vdash e_1 = e_1, e_3 \vdash e_2 = e_2, e_3 \vdash e_3 = e_3;$

$Dias_3^{15} : e_1 \dashv e_1 = e_2, e_3 \dashv e_3 = e_3, e_3 \vdash e_3 = e_3;$

$Dias_3^{16} : e_1 \dashv e_3 = e_2, e_3 \dashv e_1 = ke_2, e_1 \vdash e_1 = me_2, e_1 \vdash e_3 = ne_2, e_3 \vdash e_1 = pe_2,$
$e_3 \vdash e_3 = qe_2;$

$Dias_3^{17} : e_1 \dashv e_3 = e_2, e_1 \dashv e_2 = e_3, e_2 \dashv e_1 = e_3, e_1 \vdash e_1 = e_2 + e_3, e_1 \vdash e_2 = e_3,$
$e_2 \vdash e_1 = e_3;$

where $k, m, n, p, q \in \mathbb{C}$.

Theorem 2.3 *The derivations of three dimensional complex diassociative algebras are given as follows:*

Table 2. Derivations of three-dimensional diassociative algebras

Isom. Class	**Derivations**	**Dim. of Der.alg.**	**Isom. Class**	**Derivations**	**Dim. of Der.alg.**
$Dias_3^1$	$\begin{pmatrix} d_{11} & 0 & 0 \\ 0 & 0 & 0 \\ 0 & 0 & 0 \end{pmatrix}$	1	$Dias_3^2$	$\begin{pmatrix} d_{11} & d_{12} & 0 \\ 0 & 0 & 0 \\ 0 & 0 & 0 \end{pmatrix}$	2
$Dias_3^3$	$\begin{pmatrix} d_{11} & 0 & 0 \\ 0 & 0 & 0 \\ 0 & 0 & 0 \end{pmatrix}$	1	$Dias_3^4$	$\begin{pmatrix} d_{11} & 0 & 0 \\ d_{21} & d_{22} & 0 \\ 0 & 0 & 0 \end{pmatrix}$	3
$Dias_3^5$	$\begin{pmatrix} d_{11} & 0 & 0 \\ d_{21} & d_{22} & 0 \\ 0 & 0 & 0 \end{pmatrix}$	3	$Dias_3^6$	$\begin{pmatrix} d_{11} & 0 & 0 \\ d_{21} & d_{22} & d_{23} \\ 0 & 0 & 0 \end{pmatrix}$	4
$Dias_3^7$	$\begin{pmatrix} d_{11} & 0 & 0 \\ d_{11} & d_{22} & d_{23} \\ 0 & 0 & 0 \end{pmatrix}$	4	$Dias_3^8$	$\begin{pmatrix} d_{11} & 0 & 0 \\ d_{21} & d_{22} & d_{23} \\ 0 & 0 & 0 \end{pmatrix}$	4
$Dias_3^9$	$\begin{pmatrix} d_{11} & 0 & 0 \\ d_{21} & d_{22} & d_{23} \\ 0 & 0 & 0 \end{pmatrix}$	4	$Dias_3^{10}$,	$\begin{pmatrix} d_{11} & 0 & d_{13} \\ 0 & d_{22} & 0 \\ 0 & 0 & 0 \end{pmatrix}$	3
$Dias_3^{11}$	$\begin{pmatrix} d_{11} & 0 & d_{13} \\ 0 & d_{22} & 0 \\ 0 & 0 & 0 \end{pmatrix}$	3	$Dias_3^{12}$	$\begin{pmatrix} d_{11} & 0 & 0 \\ 0 & d_{22} & 0 \\ 0 & 0 & 0 \end{pmatrix}$	2
$Dias_3^{13}$	$\begin{pmatrix} d_{11} & 0 & d_{13} \\ 0 & d_{22} & d_{23} \\ 0 & 0 & 0 \end{pmatrix}$	4	$Dias_3^{14}$	$\begin{pmatrix} d_{11} & 0 & 0 \\ 0 & d_{22} & d_{23} \\ 0 & 0 & 0 \end{pmatrix}$	3
$Dias_3^{15}$	$\begin{pmatrix} d_{11} & 0 & 0 \\ d_{21} & d_{22} & 0 \\ 0 & 0 & 0 \end{pmatrix}$	3	$Dias_3^{17}$	$\begin{pmatrix} d_{11} & 0 & 0 \\ 0 & 3d_{11} & 0 \\ 0 & 0 & 2d_{11} \end{pmatrix}$	1
$Dias_3^{16}$	$\begin{pmatrix} 0 & 0 & 0 \\ d_{21} & 0 & d_{23} \\ 0 & 0 & 0 \end{pmatrix}$	2	$Dias_3^{16}$	$\begin{pmatrix} 0 & 0 & \beta d_{33} \\ d_{21} & 0 & d_{23} \\ 0 & 0 & d_{33} \end{pmatrix}$ $k = 0,\ \beta = p - \frac{2q}{n}$	3
	$\begin{pmatrix} 0 & 0 & d_{13} \\ d_{21} & 0 & d_{23} \\ 0 & 0 & 0 \end{pmatrix}$ $n = -p$	3		$\begin{pmatrix} d_{11} & 0 & \alpha d_{11} - 2d_{23} \\ d_{21} & d_{11} & d_{23} \\ 0 & 0 & d_{33} \end{pmatrix}$ $k, m = 0\ \alpha = p + \frac{q}{n}$	4
	$\begin{pmatrix} d_{11} & 0 & \alpha d_{11} \\ d_{21} & d_{11} & d_{23} \\ 0 & 0 & 0 \end{pmatrix}$ $m = 0,\ \alpha = p + \frac{q}{n}$	3		$\begin{pmatrix} d_{11} & 0 & \alpha d_{11} \\ d_{21} & d_{11} & d_{23} \\ \gamma d_{11} & 0 & 0 \end{pmatrix}$ $k = -1, \alpha = p + \frac{q}{n},\ \gamma = p - \frac{m}{n}$	3

where k, m, n, p, q are structure constants of the algebra $Dias_{16}^3$.

As a result of these descriptions we have

Corollary 2.1 *There is no characteristically nilpotent diassociative algebra in dimensions two and three.*

3. Acknowledgement

The research was supported by grants UniKL/IRPS/str11061, Universiti Kuala Lumpur and 05-02-12-2188RU, Universiti Putra Malaysia.

References

[1] Ancochea J M and Commpoamor R 2000 *Characteristically nilpotent Lie algebras: a survey* arXiv:1103224v1.[math.RA]

[2] Dixmier J and Lister W G 1957 Derivations of nilpotent Lie algebras *Proc. Amer. Math. Soc.* **8** pp 155-158

[3] Jacobson N 1955 A note on automorphisms and derivations of lie algebras *Proc. Amer. Math. Soc* **6** pp 281-283

[4] Khakimdjanov Yu.B., Characteristically nilpotent Lie algebras, *Math. USSR Sbornik*, **70**(1), (1990).

[5] Khakimdjanov Yu B 1990 On characteristically nilpotent Lie algebras *Soviet Math. Dokl* **40**(1)

[6] Leger G F and Togo S 1959 Characteristically nilpotent Lie algebras, *Duke Math. J.* **26** pp 623-628

[7] Loday J-L 1993 Une version non commutative dés de Lie: les de Leibniz *L'Ens. Math.* **39** pp 269-293

[8] Loday J-L, Frabetti A, Chapoton F and Goichot F 2001 *Dialgebras and Related Operads*, Lecture Notes in Math No. 1763 (Springer, Berlin)

[9] Luks E 1976 A characteristically nilpotent Lie algebra can be a derived algebra *Proc. Amer. Math. Soc* **56** pp 42-44

[10] Rakhimov I S and Al-Hossain N 2012 On derivations of some classes of Leibniz algebras *Journal Generalized Lie theory and Applications* **6** Article ID G120501 doi 10:4303/jglta/G120501

[11] Rakhimov I S and Al-Hossain N 2011 On Derivations of low-dimensional complex Leibniz algebras *JP Journal of Algebra, Number Theory and Applications* **21**(1) pp 69-81

[12] Ravisankar T S 1971 Characteristically nilpotent algebras *Canadian J. Math.* **23** pp 222-235

[13] Rikhsiboev I M, Rakhimov I S and Basri W 2010 Classification of 3-dimensional complex diassociative algebras *Malaysian Journal of Mathematical Sciences* **4**(2) pp 241-254

International Conference on Quantum Optics and Quantum Information (icQoQi) 2013 IOP Publishing
Journal of Physics: Conference Series **553** (2014) 012007 doi:10.1088/1742-6596/553/1/012007

CHAOS IN BOHMIAN MECHANICS OF COMMENSURATE HARMONIC OSCILLATORS

H Umair[1], **Z Hishamuddin**[1]
Department Physics, Universiti Putra Malaysia, Malaysia

Email: umayr.halim@yahoo.com

Abstract. Bohmian mechanics is a causal interpretation of quantum mechanics in which there are well-defined particle trajectories guided by the wave function. Within this framework, it is shown that certain classical integrable systems can exhibit both periodic and chaotic behaviours. In this paper, we extend the work of Borondo et al. by investigating the behaviours of commensurate two-dimensional harmonic oscillator systems with different ratios of frequencies. We show that the system may produce chaotic behavior of Bohmian trajectories to be dependenton frequencies ratio. This result is illustrated numerically in computer experiments displaying the Bohmian trajectories.

1. Introduction

Despite the great success of standard quantum mechanics, it still remains as an indeterministic theory, which means that it generally does not predict the outcome of any measurement with certainty [1]. This fact inspired some physicists to construct a deterministic quantum theory and one of the relatively successfull one is Bohmian mechanics.

Bohmian-type mechanics was first introduced by Louis de Broglie and independently developed again by David Bohm[2] in 1950s. Despite initial criticisms by Pauli(1953), Keller(1953), Kochen and Specker(1967) and others, this theory recently became the subject of intense theoretical and experimental study.

The critical idea in Bohmian mechanics is this theory is based on a well defined position and its trajectory guided by the wavefunction. It clearly gives advantage in term of intuitive concepts and ideas because the quantum trajectories provide causal connection between events in physical space time. Therefore this notion gives researcher a new way to solve problems in quantum mechanics. One of the problems well investigated is in the study of quantum chaos. Quantum chaos tries to understand the connection between two phenomena in physics, chaos theory and quantum theory. Note that many concepts of classical chaos (periodic orbits, attractors, phase space maps) rely on the classical notion of trajectory. However for microscopic system the situation is totally diferent because the fundamental implication of standard quantum mechanics substantially differs from classical mechanics. This is due the fact that the concept of classical mechanics like position and trajectory is not allowed because one cannot measure the position and momentum of each particle at the same time and with precision due to Heisenberg uncertainty principle. Furthermore the concept of state of physical systems in standard quantum mechanics is influenced by wave-particle duality. In terms of evolution, the wavefunction which describes the state of microscopic system obeys Schrodinger equation, and the linearity of this evolution makes the wavefunction become less sensitivity to initial condition. These factors make the notion of chaotic trajectories cannot be adopted straightforwardly in standard quantum mechanics. However physicists attempt to overcome this problem by finding quantum mechanical criteria which is able to trace the existence of chaos in the quantum domain. They found several approaches which are based on the statistical distribution of energy eigenvalue spacing and spatial correlations of wavefunction. Howeverthe fact remains that they are not directly comparable.

In recent years, Bohmian mechancis has attracted increasing attention by researchers. A number of papers discuss chaos in Bohmian mechanics for various systems, such as cat map [3], a hydrogen atom in an oscillating electric field [4], two dimensional harmonic oscillator [5] and kick rotor [6]. Chaos in Bohmian mechanics has been found when no chaos appear in standard quantum mechanics [7]. Chaos in Bohmian mechanics may also even appear when the corresponding classical system is non chaotic [8]. Furthermore several paper have shown that Lyapunov exponent which is typical chaos indicator in classical mechanics have successfully apply to demonstrated chaos in Bohmian framework. This may be demonstrated a correlation between classical and Bohmian chaos. In 1996, Guglielmo Iacomelli and Marco Pettini apply Lyapunov exponent to study the difference in stability between classical regime and quantum regime [9]. Besides that it was shown that Lyapunov exponent indicated Bohmian trajectories in one dimension will not bechaotic but maybe chaotic in the case of two dimensional [10]. Its also demonstrated that vortices generated by Bohmian trajectories can generate chaos [11]. In this respect, Bohmian mechanics may have a clear advantage to study the correlation between a quantum system and its classical counterpart because we can compare quantitatively both orbits in these two regimes using Lyapunov exponent.

This study is thus to investigate numerically the chaotic behaviours of two dimensional commensurate (rational frequency ratio) harmonic oscillator in framework of Bohmian mechanics using Lyapunov exponent. Aparticular concern is whether or not the change of rational frequency ratio will influence the behaviour of this system.

2. Bohmian mechanics formulation

In order to derive the equation describing Bohmian trajectories,let we start with Schrodinger equation

$$i\hbar\frac{\partial\psi}{\partial t}=\frac{\hbar^2}{2m}\nabla^2\psi+V(t)\psi \tag{1}$$

Now we express wavefunction in polar form.

$$\psi(x,t)=R(x,t)\exp\,[\frac{i}{\hbar}S(x,t)] \tag{2}$$

where R and S are real function. In order to get an evolution in time for R and S, we insert equation (2) into equation (1).

$$i\hbar\frac{\partial(Re^{\frac{iS}{\hbar}})}{\partial t}=\frac{\hbar^2}{2m}\nabla^2(Re^{\frac{iS}{\hbar}})+V(t)(Re^{\frac{iS}{\hbar}}) \tag{3}$$

In this way we obtain a pair of couple equation.

$$\frac{dR}{dt}=\frac{1}{2m}(R\nabla^2S+2\nabla R\nabla S) \tag{4}$$

$$\frac{dS}{dt}=-\frac{(\nabla S)^2}{2m}+V(x)-\frac{\hbar^2\nabla^2R}{2mR} \tag{5}$$

Equation (4) is the continuity equation where$R^2 = |\psi|^2$is the position probability distribution and equation (5) is identical in form to the Hamiltonian-Jacobi equation

$$-\frac{\partial S}{\partial t} = \frac{(\nabla S_c)^2}{2m} + V \tag{6}$$

However there exists two significant differences. First, Hamilton's principal function, $S_c$have the relation $\nabla S_c = p(x,t)$where p is the conjugate momenta in generalized coordinate but the term S in equation (5) represent the phase of quantum state, both term obviously express two different things. However, if we attach this physical meaning to S, $\nabla S = p(x,t)$, then we may interpret equation (5) as a modified Hamilton-Jacobi equation for real quantum particles with welldefined and trajectory. We called $\nabla S = p(x,t)$ as guidance condition where $p$is the canonical momentum of a particle in Bohmian trajectory. Second, the extra term

$$Q = \frac{\hbar^2 \nabla^2 R}{2mR} \tag{7}$$

is the only thing that makes equation (5) dffer compared to usual classical Hamiltonian-Jacobi equation. This extra term is called "quantum potential". If we attach Q with force, as a classical force is associated with a classical potential energy, then we get

$$F_Q = -\nabla Q \tag{8}$$

Here, $F_Q$is called quantum force. This expression makes Bohmian mechanics becomes more similar as Newtonian formulation, therefore the total (classical plus quantum) force become

$$F_t = -\nabla(V + Q) \tag{9}$$

Now we ready to derive the guiding equation, first we define the velocity vector field

$$v = \frac{\nabla S}{m} \tag{10}$$

as we know $\nabla S = p(x,t)$. Then by inserting equation (2) into equation (10) we finally get

$$v(t) = \frac{i}{\hbar} \frac{\psi * \nabla \psi - \psi \nabla \psi *}{|\psi|^2} \tag{11}$$

3. Two-dimensional harmonic oscillator with commensurable frequencies

The focus of this research is to study the behaviour of two-dimensional commensurate harmonic oscillator. The Hamiltonian of a particle in this system is given by

$$H(x,y) = -\frac{1}{2}\left(\frac{\partial^2}{\partial x^2} + \frac{\partial^2}{\partial y^2}\right) + \frac{1}{2}(\omega_1 x^2 + \omega_2 y^2) \tag{12}$$

where $\omega_1$and $\omega_2$are the frequencies of the oscillator, x and y are the position of the particle.

Consider the combination of eigenstates of this system

$$\psi_{0,0} = \frac{\alpha}{\sqrt{\pi}} e^{\frac{1}{2}(\omega_1 x^2 + \omega_2 y^2)} e^{-\frac{i}{2}(\omega_1 + \omega_2)t}$$
$$\psi_{0,1} = \frac{2y\alpha}{\sqrt{\pi}} e^{\frac{1}{2}(\omega_1 x^2 + \omega_2 y^2)} e^{-\frac{i}{2}(\omega_1 + 3\omega_2)t}$$
$$\psi_{1,0} = \frac{2x\alpha}{\sqrt{\pi}} e^{\frac{1}{2}(\omega_1 x^2 + \omega_2 y^2)} e^{-\frac{i}{2}(3\omega_1 + \omega_2)t}$$

Here $\alpha = (\omega_1\omega_2)^{\frac{1}{2}}$ and the eigenstate $\psi_{0,0}$ correspond to ground state energy and $\psi_{0,1}, \psi_{0,1}$ correspond to the degenerate first excited states. In this way we obtain the time evolution of the resulting wavefunction is given by

$$\left(\frac{\alpha A e^{-i(\omega_1+\omega_2)t}}{\sqrt{\pi}} + \frac{2x\alpha B e^{-\frac{i}{2}(\omega_1+3\omega_2)t}}{\sqrt{2\pi}} + \frac{2y\alpha C e^{-i(3\omega_1+\omega_2)t}}{\sqrt{2\pi}} \right) e^{\frac{1}{2}(\omega_1 x^2 + \omega_2 y^2)} \tag{13}$$

where $A = \mathrm{a} + id, B = \mathrm{b} + ig, C = \mathrm{f} + ic$. These complex numbers must satisfy normalization condition $|A| + |B| + |C| = 1$. By inserting equation (13) in equation (11), we obtain the following

$$v_x = -\frac{\sqrt{2}\alpha^2\beta_x - 2\alpha^2\gamma_x y}{V(x,y,t)} \tag{14}$$

$$v_y = -\frac{\sqrt{2}\alpha^2\beta_y - 2\alpha^2\gamma_y x}{V(x,y,t)} \tag{15}$$

where

$$\beta_x = (ab + dg)Sin(\omega_2 t) - (ag - bd)Cos(\omega_2 t)$$

$$\beta_y = (af + dg)Sin(\omega_1 t) - (ag - bd)Cos(\omega_1 t)$$

$$\gamma_x = (bf + gc)Cos(\omega_1 - \omega_2)t + (bc - gf)Cos(\omega_1 - \omega_2)t$$

$$\gamma_y = (bf + gc)Cos(\omega_1 - \omega_2)t - (bc - gf)Cos(\omega_1 - \omega_2)t$$

$$V(x,y,t) = 2x^2\alpha^2(b^2 + g^2) + 2y^2\alpha^2(c^2 + f^2) + \alpha^2(a^2 + d^2)$$

$$+2\sqrt{2}\alpha^2[(ab + dg)Cos(\omega_2 t) + (ag + bd)Sin(\omega_2 t)]x$$

$$+2\sqrt{2}\alpha^2[(af + dc)Cos(\omega_1 t) + (ac + df)Sin(\omega_1 t)]y$$

$$+4\alpha^2[(bf + gc)Cos(\omega_1 - \omega_2)t + (bc + gf)Sin(\omega_1 - \omega_2)t]xy$$

and quantum trajectories defined in Bohmian framework is the solution of this two differential form.

4. Lyapunov exponent

Lyapunov exponent provides a quantitative measure of the divergence orconvergence of nearby trajectories for a dynamical system. To do computation, Benettin et al. have introduced a procedure to calculate Lyapunov exponents for bounded systems. Consider two nearby points in phase space M with very small separation.

$$|\delta p_0| = \|x_0 - y_0\| \tag{16}$$

where x_0 and y_0 are initial point of two nearby trajectories and $\|\ldots\ldots\|$ is the Euclidean norm.Then we can define the Hamiltonian flow $\{T^t\}$where time $t \in \Re$ such that

$$x(t) = T^t x_0 \tag{17}$$

$$y(t) = T^t y_0 \tag{18}$$

Therefore the separation between these two trajectories at time t is

$$|\delta p_t| = \|x(t) - y(t)\| \tag{19}$$

In order to keep the trajectories separation within the linearized flow range, given a fixed time t we modify equations (17) and (18) by

$$x_i = T^t x_{i-1} \tag{20}$$

$$y_i = T^t y_{i-1} \tag{21}$$

where $i \in \aleph$ and equation (19) become

$$|\delta p_t| = \|T^t x_{i-1} - T^t y_{i-1}\| \tag{22}$$

Now we can define$x_1 = T^t x_0$ and$|\delta p_1| = \|T^t x_0 - T^t y_0\|$. Thereafter we have to rescale $y_i$such that$\|y_1 - x_1\| = |\delta p_0|$. Such procedure will be repeated by n times to get a sequence of positive number $\{\delta p_i\}$ and the Lyapunov exponent given by

$$\lambda = \frac{1}{t}\sum_{i=1}^{n} ln\frac{|\delta p_i|}{|\delta p_0|} \tag{23}$$

5. Results and conclusion

From the numerical calculations, we find that there exists three types of Bohmian trajectories: regular, mixture between regular and chaotic, and chaotic which are illustrated by figures 1(a), 2(a), 3(a) respectively. Figure 1(b) represent the calculation of λ for the trajectory of figure 1(a) . It shows the logarithmic value of Lyapunov number decreases when time increases, following the power law, $\lambda \sim$ 1/t. This means that the trajectory is ordered and $\lim_{t\to\infty}\lambda = 0$.Such results is also found in figure 2(b) in the early time period (t = 500), however for time larger than 500, the quantity λstabilizes at a positive value. The latter time evolution of figure 2(b) is similar to that of figure 3(b) where it shows that λdecreases slowly and tends to stabilize to a positive value, $\lambda \cong 0.1$ at t= 1000.

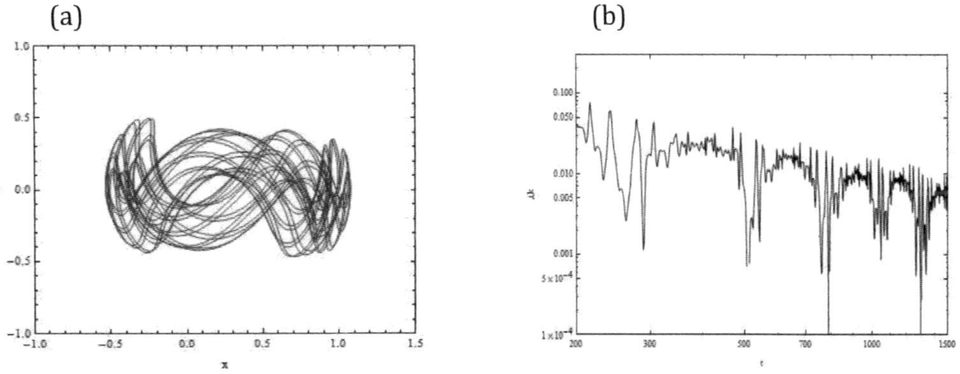

Figure 1.(a) A Bohmian orbit with initial condition x_0= 1, y_0= 0, a=0.37, b=0.44, c=0.44, d=-0.02, e=0.49, f=-0.49 for frequencies ratio 4:1. (b) The time evolution of the Lyapunov exponent

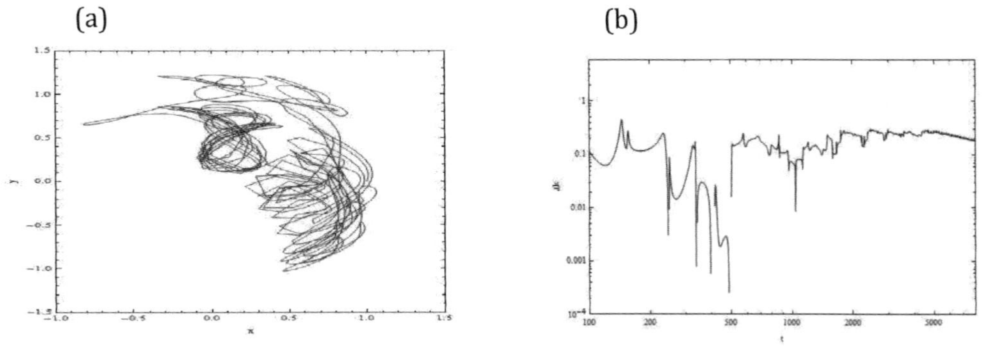

Figure 2. Same as Figure 1, for frequencies ratio 4:7.

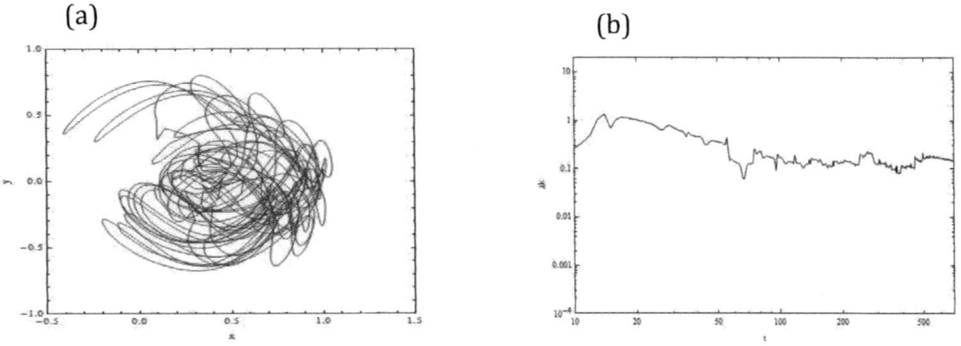

Figure 3. Same as Figure 1, for frequencies ratio 7:3.

From these results, we can conclude that a system which is classically integrable (commensurate two dimensional harmonic oscillators) may have three types of trajectories: regular, mixture between regular and chaotic, and chaotic dependent on the frequency ratio.

6. Acknowledgement
I would like to express gratitude to my supervisor, Associate ProfessorDr. Hishamuddin Zainuddin for his invaluable support, encouragements, supervisionthroughout this research work. I would also like to thank to Kementerian Pengajian Tinggi for give me scholarship along my study leave.

7. References
[1] JohnFD2009*Quantum Mechanics: Fundamental Principles and Application* (Durham: University of New Hampshire)
[2] Bohm D 1952 *Phys. Rev.* **85** 166
[3] Makri N 2004 *J. Phys. Chem.* A **108** 806
[4] Gindensperger E, Meyer C and Beswick J A 2002 *J. Chem. Phys.* **116** 8
[5] Prezhdo O V and Brooksby C 2001 *Phys. Rev. Lett.* **86** 3215
[6] BerryMV 1977*J. Phys.* A **10** 2083
[7] Heller EJ 1984 *Phys. Rev. Lett.***53** 1515
[8] Faisal F H Mand Schwengelbeck U 1995 *Phys. Lett.* A**207** 31.
[9] Iacomelli G and Pettini M 1996*Phys. Lett.* A **212** 29.
[10] Efthymiopoulos C and Contopoulos G 2006 *J. Phys. A: Math. Gen.* **39** 1819
[11] Zheng Y 2008*Chaos,Solitons & Fractals***36**263-270
[12] Bonfim D A, FlorencioJ, Sá Barreto FC 1998 *Phys. Rev. E***58** 2693
[13] Parmenter RH and Valentine RW 1995 *Phys. Lett. A***201** 1
[14] Wu H and Sprung D W L *Phys. Lett. A*1994 **196** 229.
[15] Wisniacki D A and PujalsE R*Europhys. Lett* 2005 **71** 159–165

Journal of Physics: Conference Series **553** (2014) 012008 doi:10.1088/1742-6596/553/1/012008

An Analysis of the Spekkens Toy Theory with Connection to Wootters Discrete Phase Space

Mojtaba Aliakbarzadeh[1] **and Hishamuddin Zainuddin**[1,2]

[1] Laboratory of Computational Science and Mathematical Physics, Institute for Mathematical Research, Universiti Putra Malaysia, 43400 UPM Serdang, Selangor, Malaysia

[2] Department of Physics, Faculty of Science, Universiti Putra Malaysia, 43400 UPM Serdang, Selangor, Malaysia

E-mail: `mojtabaalek@gmail.com`

Abstract.

The toy model of Spekkens is a formalism which can partially describe quantum mechanics. The theory deals with the (epistemic) states of a spin-1/2 particle, or qubits and it is closely related to the discrete phase space formalism of Wootters and collaborators. One can apply the stabilizer formalism for finding similarities of these two models. Noting that MUB basis vectors are obtained by eigenstates of generalized Pauli operators, the MUB basis vectors are thus the set of stabilizer states. Galvao has characterized the set of states with non-negative Wigner function class; they form the convex hull of the stabilizer states used as the MUB basis vectors. By combining both approaches, one can show epistemic states that are analogous to the convex hull of the stabilizer states (used as basis vectors in the MUB set) always make valid nonmaximal knowledge epistemic states.

1. Introduction

The Wigner function plays the role of a quasi-probability distribution on continuous phase space. While this real-valued function has many properties of probability distributions, it can take negative values. This negativity can be considered as nonclassical features of such quantum states [6]. In the same way as the original Wigner function, Wootters et. al [2, 3, 4] constructed Wigner functions on discrete phase spaces to describe finite-dimensional quantum systems. For determining this discrete phase space, they labelled the axes of the phase space with finite field (Galois field) of N elements where N is power of prime and at the same time with the quantum states of the system.

The discrete phase space of Wootters in d-dimensional Hilbert space is a $d \times d$ real array and consists of lines as a set of points. It can be divided to parallel lines which are called striations. Parallel lines in each striation may not be really similar to the geometrical image of parallel lines in Euclidean space, but they have the similar feature that they never intersect with each other at any point. It should be mentioned that non-parallel lines at two different striations intersect at just one point. This feature is the result of using the Galois field to form p and q axes of the discrete phase space. For constructing discrete Wigner function Wootters et al. associate vectors of each basis with specific lines of the corresponding striation. This association is arbitrary and each separate choice leads to different class of Wigner function.

There is great interest to characterize quantum states with classical properties in the sense of giving rise to non-negative discrete Wigner function [5, 7]. On the other hand, while Spekkens toy model [1] is impotent to reproduce the violation of Bell inequalities and the existence of a Kochen-Specker theorem, we want to relate its states to the subset of quantum states which has non-negative Wigner functions. Spekkens toy model is based on foundational principle that there is a balance between knowledge and ignorance in a state of maximal knowledge From this starting point, and a few other assumptions, one can derive the toy theory.

In this paper we try to implement the ontic and epistemic states of the Spekkens model in the discrete phase space formalism. We organize our approach for one and two qubits in following sections.

2. Elementary systems

In Spekkens toy model disjunction of four ontic states "1" , "2", "3" and "4" lead to six epistemic states which can be represented graphically as follows (symbol $\bigvee$ uses as notation of disjunction)

$1 \vee 2 \Longleftrightarrow |0\rangle \Longleftrightarrow$  $2 \vee 4 \Longleftrightarrow |-\rangle \Longleftrightarrow$ 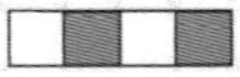

$3 \vee 4 \Longleftrightarrow |1\rangle \Longleftrightarrow$  $2 \vee 3 \Longleftrightarrow |+i\rangle \Longleftrightarrow$ 

$1 \vee 3 \Longleftrightarrow |+\rangle \Longleftrightarrow$ 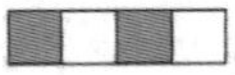 $1 \vee 4 \Longleftrightarrow |-i\rangle \Longleftrightarrow$ 

as it can be seen these six epistemic states are analogous to the six pure qubit states indicated.

In the case of discrete phase space for a single qubit (which is equivalent to a single spin-1/2 particle), Wootters et al. [4] associate the horizontal axes of phase space to eigenstates of σ_z i.e. states $|\uparrow\rangle$ and $|\downarrow\rangle$, and the vertical axes to eigenstates of σ_x i.e. states $|\leftarrow\rangle$ and $|\rightarrow\rangle$

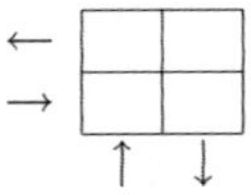

The two choices for the eigenstates of σ_y are equally natural. Note that, in the case of more qubits, there are multiple ways of making these associations which will lead to different Wigner functions for the same fixed set of mutually unbiased basis (MUBs) [2, 3, 4].

Comparing discrete phase space with the epistemic states of a qubit, we can have following associations:

$1 \vee 2 \Longleftrightarrow |0\rangle \Longleftrightarrow$  $2 \vee 4 \Longleftrightarrow |-\rangle \Longleftrightarrow$ 

$3 \vee 4 \Longleftrightarrow |1\rangle \Longleftrightarrow$ 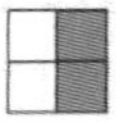 $2 \vee 3 \Longleftrightarrow |+i\rangle \Longleftrightarrow$ 

$1 \vee 3 \Longleftrightarrow |+\rangle \Longleftrightarrow$  $1 \vee 4 \Longleftrightarrow |-i\rangle \Longleftrightarrow$ 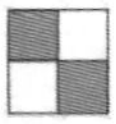

If we name the four ontic states of Spekkens model as:

1	2	3	4

it should correspond to the four states of Wootters model

$\leftarrow$	2	4
$\rightarrow$	1	3
	$\uparrow$	$\downarrow$

We can conclude ontic state "1" relates to state $\uparrow\rightarrow$, "2" to $\uparrow\leftarrow$, "3" to $\downarrow\rightarrow$ and "4" to $\downarrow\leftarrow$, so we can assume states $\uparrow\rightarrow$, $\uparrow\leftarrow$, $\downarrow\rightarrow$ and $\downarrow\leftarrow$as ontic states of Spekkens model.

1	2	3	4
$\uparrow\rightarrow$	$\uparrow\leftarrow$	$\downarrow\rightarrow$	$\downarrow\leftarrow$

3. Pairs of elementary systems

Extending this process for two-qubit case, we obtain the following picture where the rows represent the different ontic states of the first system (identified by subscript 1) and the columns represent the different ontic states of second systems (identified by subscript 2).

$(\downarrow\leftarrow)_1$				
$(\downarrow\rightarrow)_1$				
$(\uparrow\leftarrow)_1$				
$(\uparrow\rightarrow)_1$				
	$(\uparrow\rightarrow)_2$	$(\uparrow\leftarrow)_2$	$(\downarrow\rightarrow)_2$	$(\downarrow\leftarrow)_2$

The phase space for 2 qubits has $N = 2^2 = 4$ dimensions which brings about a 4×4 array of points for its phase space. Wootters associated the axes of the phase space by the discrete variables p and q, which take values in $GF(N) = GF(4)$ and generated lines and striations in this discrete phase space labeled by the Galois field elements. Also conceiving qubits as spin states of spin-1/2 particle, he associated the horizontal axis of the phase space by the (composite) states $|\uparrow\uparrow\rangle$, $|\uparrow\downarrow\rangle$, $|\downarrow\uparrow\rangle$, $|\downarrow\downarrow\rangle$, and the vertical axis by the states $|\rightarrow\rightarrow\rangle$,$|\rightarrow\leftarrow\rangle$, $|\leftarrow\rightarrow\rangle$, $|\leftarrow\leftarrow\rangle$ as shown here

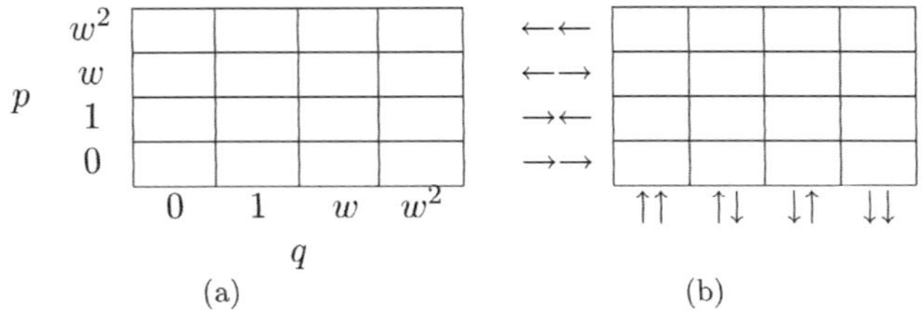

Figure 1: A 4×4 array of points of phase space. The axis of phase space associated (a) by the elements of $GF(N)$ and (b) spin states.

Comparing these two platforms, we can directly obtain following relation between cells of them, which can be used in inverting a state from one model to the other.

Journal of Physics: Conference Series **553** (2014) 012008 doi:10.1088/1742-6596/553/1/012008

$(\downarrow\leftarrow)_1$	11	15	12	16
$(\downarrow\rightarrow)_1$	3	7	4	8
$(\uparrow\leftarrow)_1$	9	13	10	14
$(\uparrow\rightarrow)_1$	1	5	2	6
	$(\uparrow\rightarrow)_2$	$(\uparrow\leftarrow)_2$	$(\downarrow\rightarrow)_2$	$(\downarrow\leftarrow)_2$

$\leftarrow\leftarrow$	13	14	15	16
$\leftarrow\rightarrow$	9	10	11	12
$\rightarrow\leftarrow$	5	6	7	8
$\rightarrow\rightarrow$	1	2	3	4
	$\uparrow\uparrow$	$\uparrow\downarrow$	$\downarrow\uparrow$	$\downarrow\downarrow$

Using this relation one can convert some exemplary states from Spekkens model to discrete phase space model; for instance, the first line of the following picture indicates some states in Spekkens model and the second line are their equivalent states in discrete phase space for uncorrelated composite systems.

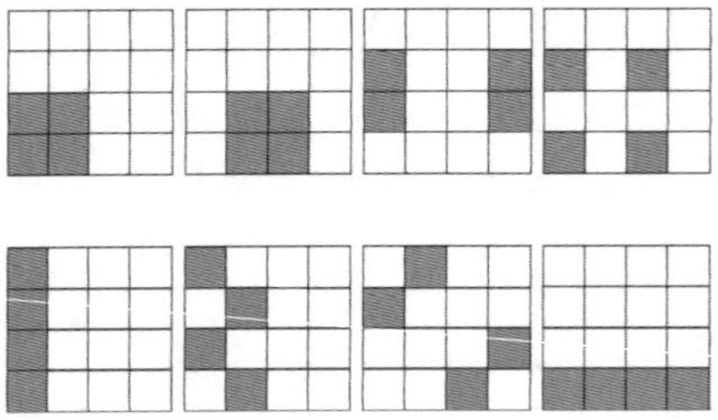

For correlated composite systems the equivalent states in discrete phase space are as the second line of following picture.

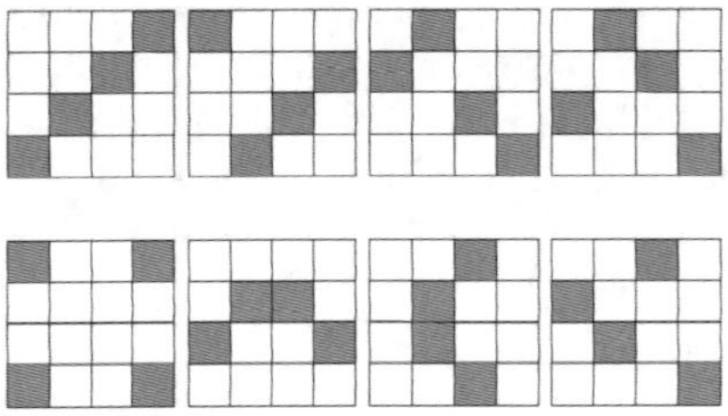

This relation also can be applied to convert the 5 striations of discrete phase space for 2 qubits to Spekkens epistemic states. Note that there are multiple ways of making associations between lines of striations and states of a fixed MUB. In general, this will lead to different definitions of Wigner function. But here we don't care about these different possibility of Wigner function. Actually we assumed lines of striations as the representative of states of fixed MUB, then we try to find their relation with epistemic states. It is clear that the actual accordance is between epistemic states and MUB vectors, therefore to calculate Wigner function for epistemic states, there are again multiple ways of making associations between lines of striations and states of epistemic toy model.

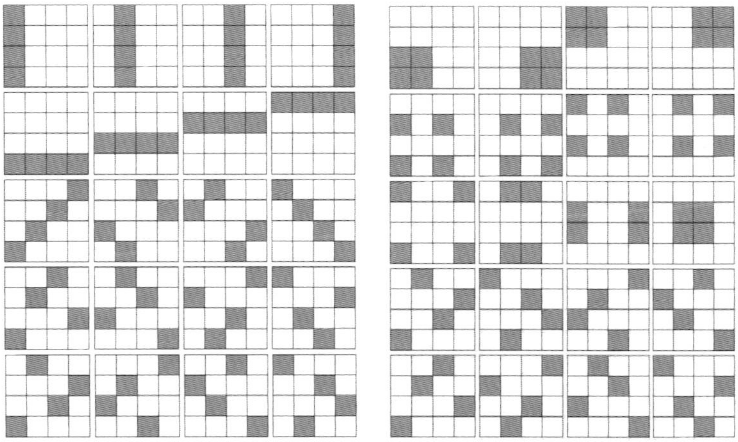

Figure 2: Converting (a) the 5 striations of discrete phase space for 2 qubits to (b) Spekkens epistemic states.

It is worth mentioning that, these five bases of epistemic states completely matches the sets of five mutually unbiased partitionings (MUPs) that Spekkens defined in his paper [1]. Similar to the set of MUBs for two qubits which has three fully separable bases and two non-separable ones [8], here also we have three sets of uncorrelated states and two sets of correlated states. But these MUP states don't cover all the possible correlated and uncorrelated states (there are some correlated and uncorrelated states which are not in this set) for example like

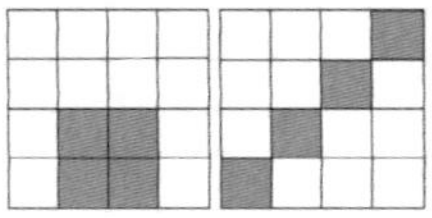

which are analogous to states $|0\rangle|+i\rangle$ and $\frac{1}{\sqrt{2}}(|00\rangle+|11\rangle)$ respectively but they don't belong to the MUB bases vectors. For finding these analogue states we apply the method of Matthew Pusey [9] where he suggested a way to compare the qubit stabilizer formalism and Spekkens toy theory. He introduced the toy stabilizer theory which can be mapped to and compared with qubit stabilizer theory. Using this innovative stabilizer notation he listed analogous states to Spekkens states.

We can see one of the interesting results of the comparison between the epistemic toy model and discrete phase space is in the case of epistemic states of nonmaximal knowledge. The states of nonmaximal knowledge in the Spekkens model are actually mixed states and they can be considered as convex combination of pure states. Spekkens gave some examples of epitemic states of nonmaximal knowledge that violate the principle in some way

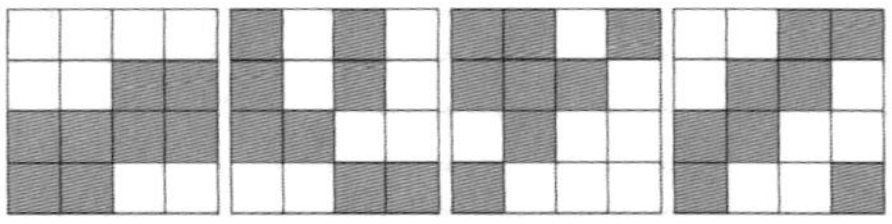

We found that the convex combination of MUP states always make valid nonmaximal-knowledge epistemic states. It is consistent with Galvao et. al [5, 7] theorem that the convex hull of MUB basis vectors leads to non-negative discrete Wigner function; they defined the set of pure and mixed states C_d for d-dimensional Hilbert space having non-negative discrete Wigner functions. They claim that the only pure states having non-negative Wigner functions are stabilizer states, it is clear the only pure states in C_d are the MUB projectors which are stabilizer states (simultaneous eigenstates of Pauli operators). One should consider that those epistemic states out of MUP are similar to Pure stabilizer states out of C_d which lead to non negative Wigner function.

Acknowledgments This work is carried out under the ERGS grant 1-2013-552718, Ministry of Education, Malaysia.

References

[1] Spekkens R W 2007 Evidence for the epistemic view of quantum states: A toy theory *Physical Review* **A 75(3)** 032110
[2] Wootters W K 1987 A Wigner-function formulation of finite-state quantum mechanics *Annals of Physics* **176(1)** 1-21
[3] Wootters W K 2004 Picturing qubits in phase space *IBM Journal of Research and Development* **48(1)** 99-110
[4] Gibbons K S, Hoffman M J and Wootters W K 2004 Discrete phase space based on finite fields *Physical Review* **A 70(6)** 062101
[5] Galvao E F 2005 Discrete Wigner functions and quantum computational speedup *Physical Review* **A 71(4)** 042302
[6] Kenfack A and yczkowski K 2004 Negativity of the Wigner function as an indicator of non-classicality *Journal of Optics* **B**: Quantum and Semiclassical Optics **6(10)** 396
[7] Cormick C, Galvao E F, Gottesman D, Paz J P and Pittenger A O 2006 Classicality in discrete Wigner functions *Physical Review* **A 73(1)** 012301
[8] Romero J L, Bjork G and Klimov A B 2005 Structure of the sets of mutually unbiased bases for N qubits *Physical Review* **A 72(6)** 062310
[9] Pusey M F 2012 Stabilizer notation for Spekkens toy theory. *Foundations of Physics* **42(5)** 688-708

Nitrogen Effect on the Frequency Shift in Indium-GaAsN Quantum Wells Laser

A.Ouerdane[1, 2], M.Bouslama[2], A. Abdellaoui[2], M.Ghaffour[2] and Y. Al-Douri[3]
[1] Université de Khemis Miliana route de Theniet El Had 44225 Ain Defla, Algeria
[2]Laboratoire Matériaux LabMat ENPO Mnaouer BP 1523 Oran 31000, Algeria
[3] Institute of Nanoelectronic Engineering, University Malaysia Perlis, Malaysia

Email: ouerdanea@yahoo.fr

Abstract. The GaAsN/GaAs material system has recently been investigated as material for the realization of high-performance laser diodes emitting at the optical fiber window. Nitrogen content in GaAsN has a great influence on the band gap in this semiconductor. The knowledge of influence of nitrogen on optical transition energies in GaAsN/GaAs Quantum Wells (QWs) or Quantum Dots (QDs) structures is very important due to the possibility of adjustment of the optical transition energies to the telecommunication wavelength range when the system GaAsN is submitted to further indium elements giving rise to the quaternary alloys (In)GaAsN. The purpose of this work is to solve the coupled nonlinear rate equations describing the complex electric field and the carrier density in a simple model of the GaN_xAs_{1-x} or (In)GaAsN semiconductor laser. The model is sufficient to account for many of the observed dynamics in a single mode semiconductor laser in response to a dynamic drive current, such as relaxation oscillations and frequency chirping.

1. Introduction

A semiconductor laser consists of an active region sandwiched between a p-n junctions made of a direct band gap material, the n-side contains an excess of electrons, while the p-side contains an excess of holes. Through this junction there is formed a potential barrier which prevents recombination between electrons and holes. Direct polarization of the junction reduces the barrier and facilitates the recombination of the carriers. If the forward bias is large, there is a population inversion which relates the gain required for lasing [1-3].The interaction between the active region and the laser light is maximized by the use of a double heterostructure constituted by a combination of materials with different band gaps. This structure ensures the confinement of photons and charge carriers due to differences in refractive index and forbidden band widths, respectively. A low concentration of nitrogen in binary GaAs lattice generates a dramatic effect on electronic structure and optical properties of this compound. By replacing weak amounts (1%) of arsenic by nitrogen gives a ternary compound GaNAs and lead to reduce the band gap that reaches about 0.2 eV. GaN_xAs_{1-x} ,with "x" represented the proportion of nitrogen replacing the As element, is a very interest compound used as active region of semiconductor diode laser giving rise to a wavelengths laser between 1330 nm and 1555 nm which correspond to the optical fiber window very useful for optical telecommunications. [4-7].The optical power emitted by the diode laser is proportional to the current supplied to the GaN_xAs_{1-x} semiconductor heterojunction. The laser dynamics can be modeled by coupled rate equations which describe the relation between the carrier number, the photon density *and* the optical phase [4]. The purpose of this work is to solve the coupled nonlinear rate equations describing the complex electric field and the carrier density in a simple model of the GaN_xAs_{1-x} semiconductor laser. The model is sufficient to account for many of the observed dynamic such as relaxation oscillations and frequency chirping [8, 9].

2. Experimental

Trimethyl gallium arsine AsH_3 and dimethylhydrazine (DMHy) were used to grow GaNAs of thickness 0.2μm on GaAs substrates. The vector gas N_2+H_2 with the ratio of 80-20% respectively. The compound surface GaNAs are characterized by AES and EELS using a hemispherical analyzer operating in direct mode N (E). Owing to the Auger Electron Spectroscopy; we give some results related to the effect of the temperature on GaAsN heated during its preparation process. The Auger signals N-KLL has been recorded as shown on fig. 1. Through the intensity of both peaks, we note that GaAsN heated at 730°C involves less nitrogen in comparison to GaAsN no heated. Hence, we conclude that the composition of nitrogen decreases as function of the temperature. This might be due to desorption of excess nitrogen from the GaAsN surface [10]. This heating process leads to give the GaAsN/GaAs stable. Such an aspect is confirmed by the Electron Energy Loss Spectroscopy (EELS) during the irradiation process by the electron beam as shown on fig.2.Fig.3a illustrate the variation of the energy band of a quantum well based (In) GaAsN showing the geant optical mode in the InGaAsN semiconductor laser cavity.

The analyzis of our samples by spectroscopic techniques such as AES and EELS aims to the realization of quantum wells .These samples will be then charactirized by photoluminescence PL to determine the emission wavelength and energy levels. Each quantum well is in itself a laser diode as shown in fig.3b. Waiting to perform this experiment by designing quantum wells and the realisation of ohmic contact we began first with a simulated study. The experimental investigation of QWs laser diodes will be published after.

3. Simulation of the diode laser GaNAs.

3.1 Band gap variation

The band gap of the relaxed unstrained alloy is known as:

$$Eg(In_xGa_{1-x}N) = x\,Eg(InN)+(1-x)\,Eg(GaN) - bx(1-x) \qquad (1)$$

Where x is the proportion of Nitrogen injected into the alloy (In)GaAsN which compensate the (1-x) of arsenium of (In)GaAsN alloy, Eg is the energy band gap and "b" is the bowing parameter. Low Nitrogen content demonstrate large band gap bowing with only few % Nitrogen. Typical bowing coefficient for semiconductor alloys are on order of 0.1 eV and are fairly sensitive to the composition. However, bowing coefficients for GaAsN and (In)GaAsN alloys is considerably augmented, reached the values of 10eV and vary strongly with composition. Whence the band gap reduction is possible with 1% Nitrogen content or less as shown in Fig.4a. Fig.4b is related to the (In)GaAsN refractive index as function of the wavelengths and the composition x of indium. The best applications for these III-V materials are the lasers with wavelength in the 1.33 to 1.55 range that match with optic fiber windows.

Journal of Physics: Conference Series **553** (2014) 012009 doi:10.1088/1742-6596/553/1/012009

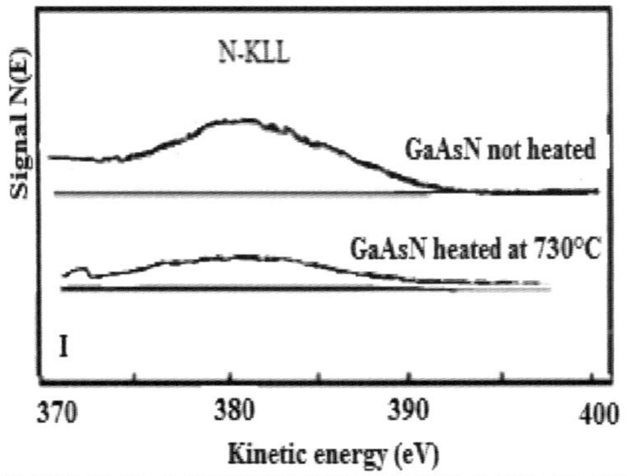

Figure.1:AES N-KLL spectra of GaAsN compound before and after heating

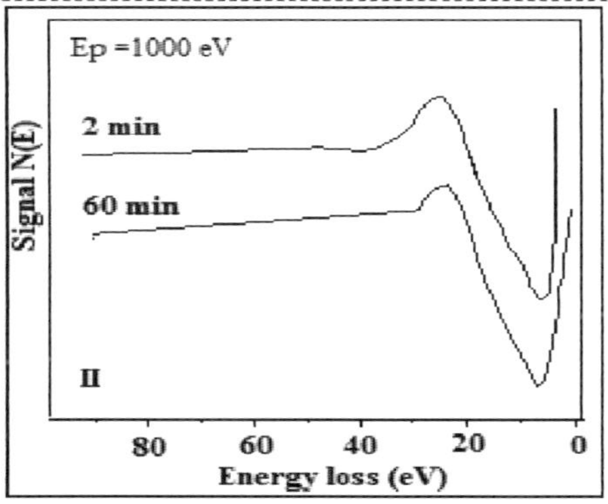

Figure.2: EELS spectra of GaNAs heating at 730°C for different exposure time.

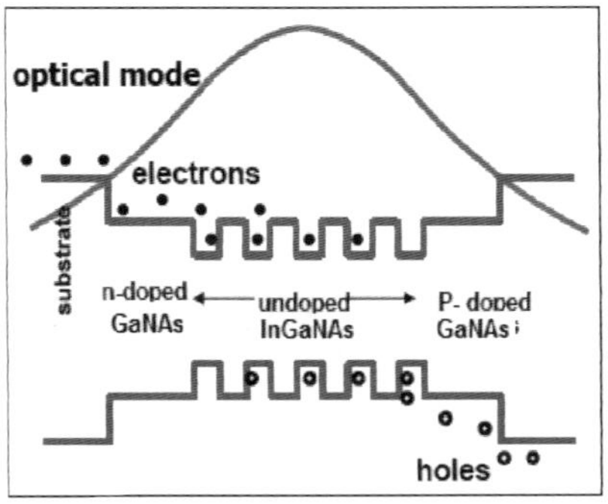

Figure.3a: current injection creating circulation of electrons and holes and generating inversion population giving rise to an optical laser mode

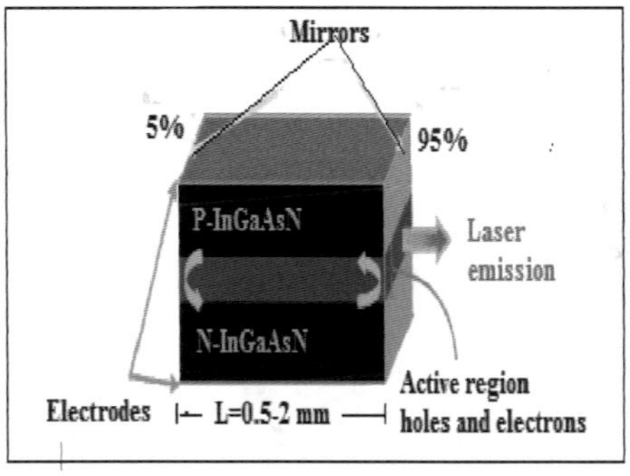

Figure3b: Diode laser conception showing the active region , the P-InGaAsN, the N-InGaAsN and the two cavity mirrors

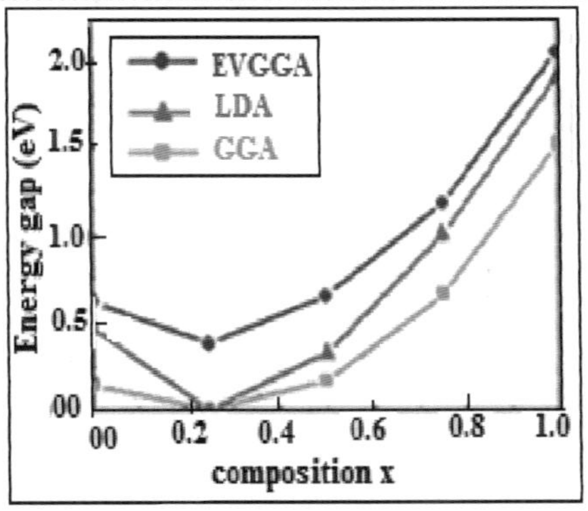

Figure.4a Simulated variation of GaAsN with the nitrogen concentration

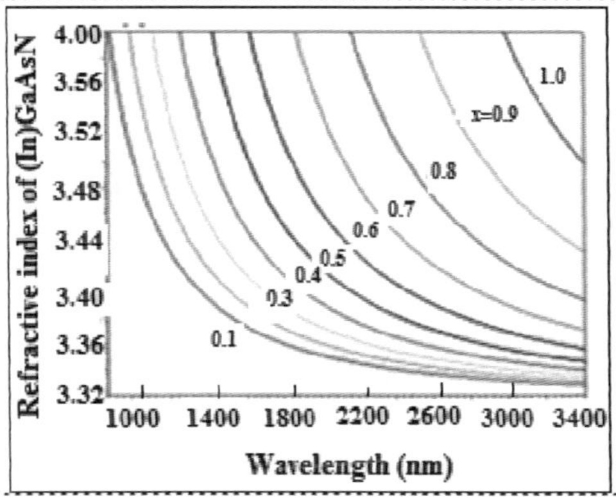

Figure.4b: (In)GaAsN refractive index as function of the wavelengths.

The optical power emitted by the diode laser is proportional to the current supplied to the GaNAs semiconductor heterojunction. The laser dynamics can be modeled by coupled rate equations which describe the relation between the carrier number, the photon density and the optical phase [14]. To use these models, values for the rate equations, parameters must be chosen appropriately in order to obtain agreement between simulated and measured results for system performance [1].

The purpose of this work is to solve the coupled nonlinear rate equations describing the complex electric field and the carrier density in a simple model of the semiconductor laser. The model is sufficient to account for many of the observed dynamics in a single mode semiconductor laser in response to a dynamic drive current, such as relaxation oscillations and frequency chirping [1, 15]. The output of the solver is the time, injection current, and the transformed components of the state vector, computed at discrete time steps. The rate equations are integrated using the fourth order Runge-Kutta computation [1, 16].

3.2 Rate equations: steady state

The laser dynamics can be modeled by coupled rate equations which describe the relation between the carrier number Np (t), the photon density $S_p(t)$ and the optical phase $\varphi(t)$. [1-6]

$$\frac{dN_p(t)}{dt} = \frac{I(t)}{q} - \frac{N_p(t)}{\tau_n} - g(N,T)\,\frac{N_p(t) - N_0}{1+\varepsilon S_p(t)} S_p(t) \tag{2}$$

$$\frac{dS_p(t)}{dt} = \Gamma\, g(N,T)\frac{N_p(t) - N_0}{1+\varepsilon S_p(t)} S_p(t) - \frac{S_p(t)}{\tau_p} + \Gamma\frac{\beta N_p(t)}{\tau_n} \tag{3}$$

$$\frac{d\varphi(t}{dt} = \frac{\alpha_o}{2} g(N,T)[N_P(t) - N_0\,] \tag{4}$$

N_0 is the carrier number at transparency, τ_p is the photon lifetime, τ_n is the carrier lifetime, Γ is the optical confinement factor describing the confinement mode in the active region, β is the spontaneous

$$P(t) = \frac{\eta h\upsilon}{\tau_P} S_P(t) \quad \text{and} \quad I_{th} = \frac{q}{\tau_n}\left\{n_0 + \frac{1}{g_0\tau_p}\right\} \tag{5}$$

emission factor, ε is the gain compression factor, g(N,T) is the optical gain coefficient dependent on the carrier density and the temperature function, I(t) is the injected current, α_o is the linewidth enhancement factor and q is the electron charge. The output power and the threshold current are given by: [1]. Where η is the quantum efficiency, h is the Planck constant and υ is the radiation frequency.

The steady-state solution to the rate equations is to be obtained by setting all the time derivatives $\frac{dN_p(t)}{dt}$ and $\frac{dS_p(t)}{dt}$ to zero. The carrier concentration that satisfies a given steady-state injected current is obtained by iterative self-consistent solutions of the two coupled equations. Hence, the couple rate equations (1) and (2) will be written as follow:

$$\frac{I(t)}{q} - \frac{N_p(t)}{\tau_n} - g(N,T)\frac{N_p(t) - N_0}{1+\varepsilon S_p(t)} S_p(t) = 0 \tag{6}$$

$$\Gamma g(N,T)\frac{N_p(t) - N_0}{1+\varepsilon S_p(t)} S_p(t) - \frac{S_p(t)}{\tau_p} + \Gamma\frac{\beta N_p(t)}{\tau_n} = 0 \tag{7}$$

The carrier number N will be deduced as follow:

$$N = \frac{\frac{I(t)}{q.V} - \frac{\tau_n S_p}{\tau_p \Gamma}}{\beta - 1} \tag{8}$$

4. Photon density S(f) and Frequency Response of the Semiconductor Laser.

For laser with small signal, the expression of the photon density S(f) and transfer function *R(f)* above threshold are as follow[1,15].

$$R(f) = \frac{S_p(0)}{i\,(0)} = \frac{f_0^2}{f_0^2 - f^2 + jf f_d} \tag{9}$$

$$f_0 = \frac{1}{2\pi}\sqrt{\frac{g_0\eta}{\tau_p}(I - I_{th})} \tag{10}$$

$$f_d = \frac{\epsilon\,\eta(I - I_{th})}{2\pi\tau_p}, \tag{11}$$

$$f_p = \sqrt{f_0^2 - \frac{f_d^2}{4}} \tag{12}$$

Where R (f), f_0, f_d and fp are respectively the frequency response, the main frequency, the damping frequency and the frequency of peak response.

5. Numerical Simulation and Results.

5.1 Steady state solution

A desired modulated laser signal can be generated by direct injection of the laser diode by an electric signal The carrier number Np *(t),* the photon density or optical signal Sp*(t*) will carry the same information as the modulating electric signal I(t) does. Therefore, in the absence of current modulation, the laser diode emits a constant-power monochromatic light. The current modulation induces a variation of the power and frequency of the beam.

The deriving current of a modulated laser diode is the sum of two terms. [16]. Where is a time function format witch describe the signal current. The first term represents the bias current, I_b, which sets a value above the threshold level of the laser while the second term, which determines the modulation level, is noted I_m. Both currents are adjusted to achieve the desired variable power.

In fig.5I and fig.5II we can see the simulated I(t) and the laser signal for different wavelength known in optical telecommunication using optical fiber. It becomes clear that the optical signal can carry the same information as the modulating electric signal does.

Fig.6a and fig.6b highlight the contribution of nitrogen on the displacement of the laser wavelength towards the red and shows a bandwidth narrower and more damped when the wavelength increases. Hence as cited above, the introduction of weak amount of nitrogen in GaAs compound allows great flexibility in terms of band gap energy and stress state. [19-25]

5.2. Amplitude modulation and frequency response

To obtain transient solutions to the rate equations we have to solve numerically the rate equations to obtain the optical power and the phase of the electrical field at the laser output. For this, we use a fourth-order Runge-Kutta algorithm; we proceed first by a number of simplifying assumptions such as:

(1) The photon and electron distributions are spatially uniform

(2) The refraction index is spatially uniform and the effect of its variations with time is neglected,

(3) The optical confinement factor and the spontaneous emission factor are treated as constant. [1,12].

It will be possible to obtain very short optical pulses, suitable for high-speed light wave telecommunication systems, from a quasi-rectangular modulation current as shown in fig.7. In practice, optical modulation is a way to minimize the effects of electrical parasitics. Within the model, this is equivalent to adding an optical perturbation to optical signal generated by the carriers. The frequency response of the semiconductor laser with the frequency response of the mount fixture is plotted on fig.8. We note that the bandwidth is mainly controlled by the resonance frequency. The damping due to the injected current variation is simulated and recorded, the latter being significantly influenced by parasitics and adiabatic phenomenon. We note also that the bandwidth increases when the bias current increases [1, 18].

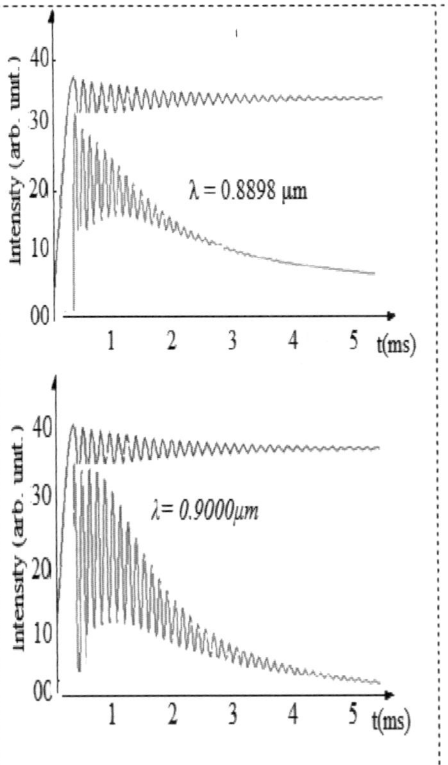

Figure.5.I: modulated laser derived by the current source I(t) = 40 mA for two wavelength λ=0.8898 μm and λ=0.9000 μm corresponding to InGaAs compound.

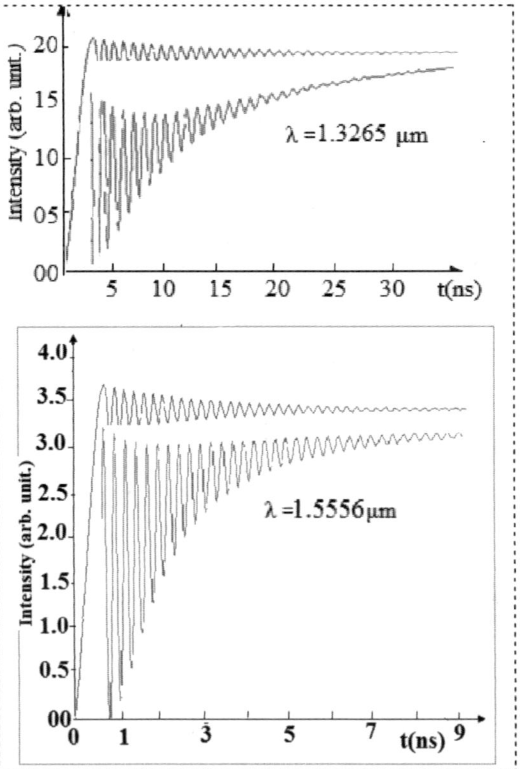

Figure.5.II: modulated laser derived by the current source I(t) =30 mA for two wavelength λ=1.3265μm (GaNAs) and λ=1.5540 μm (InGaNAs)

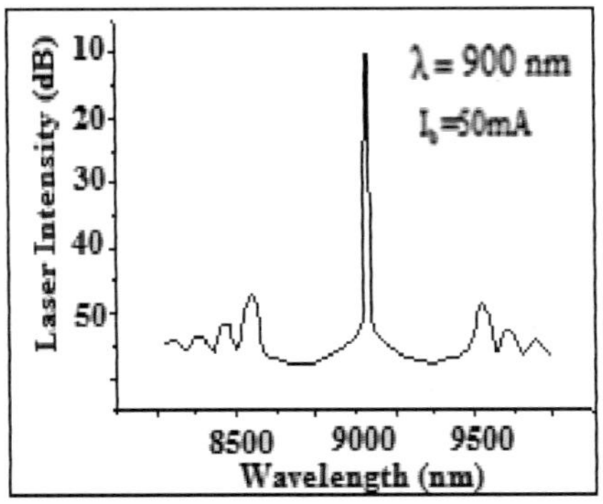

Figure.6a: (In)GaNAs where nitrogen atom lead to a spectrum centered on λ=9000 nm Δλ =5 nm

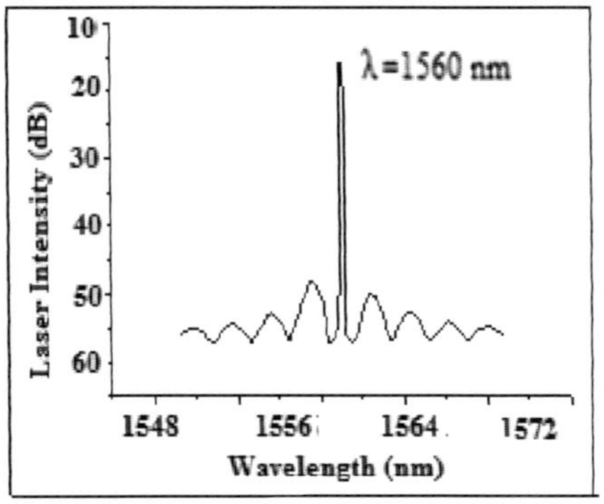

Figure.6b: λ=1560 nm with a bandwidth Δλ =1nm with less than 1% of Indium.

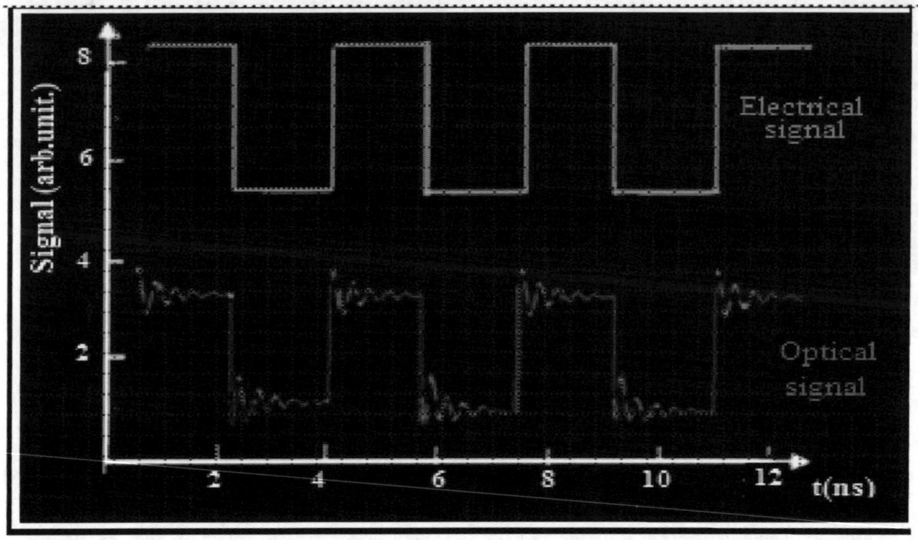

Figure 7.: Transient dynamics of carrier density and photon number for high-speed light wave telecommunication system

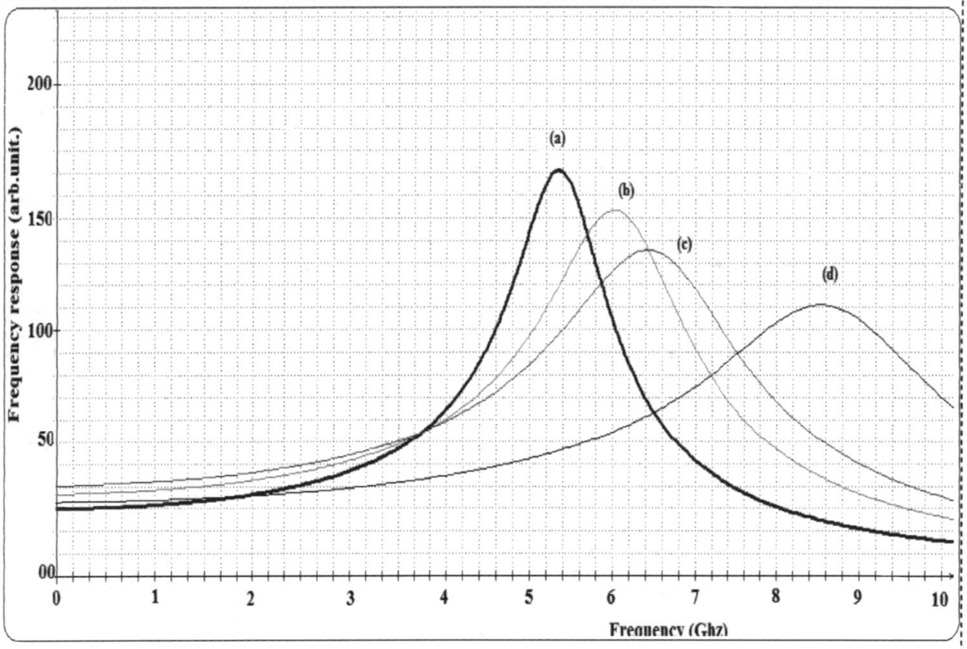

Figure.8 : Variation of the frequency response intensity and the bandwith with the bias current I_b : a) I_b=30μA ; (b) I_b=40μA, (c) I_b=50μA, (d)) I_b=60μA

6. Conclusion

We have study the material GaN_xAs1_{-x} with sensitive spectroscopic methods as AES and EELS. We considered it as active region of a diode laser. The quantum well structure giving rise to diode lasers has been studied by varying the concentration in Nitrogen and by modulating the response of laser diode which emission is being possible in different wavelengths following the concentration of the Nitrogen. The simulation program developed can be used to illustrate the performance of a waveguide laser diode as function of device parameters for all semiconductors electrically pumped laser diode. Adding of weak amount of nitrogen in GaAs compound allows great flexibility in terms of band gap energy, appropriated wavelengths for telecommunication systems and stress state visibility.

7. References

[1] A.Ouerdane, M.Bouslama, A. Abdellaoui, M.Ghaffour and Y. Al Douri 2013 Journal of Physics: Conference Series **435** 012028
[2] J. lloyd Hughes, E. Castro-Camus, M.B. Johnston 2005 Solid State Communication **136**.595-600
[3] Virginie Moreau 2009 Thèse de Doctorat Ecole Doctorale Sciences et Technologie de l'Information des Télécommunications et des Systèmes Université Paris-Sud11 Orsay France
[4]M.Geddo, M.Patrini, G.Guizzetti,M.Galli, R.Trotta, A.Polimeni, M.Capizzi,F.Martelli, and S.Rubini 2011 Journal of Applied Physics **109** 123511
[5] Narjis Begum 2010 PhD CIIT/SP05_PPH-005/ISB COMSATS Institute of Information Technology Pakistan
[6] F. Maskuriy, M. S. Alias, S.M. Manaf 2011 Journal of Applied Optoelectronics and Advanced Materials **13** 1213
[7] Arkadiusz Mika et al. 2013 Optica Applicata, **XLIII, 1** 53-60
[8] Ya Yan Lu Numerical Method For Differential equations Department of Mathematic City University of Hon Gong , site: http://math.cityu.edu.hk/~mayylu/ma3514/3514.pdf
[9] S.Habermann, J.Hofeldt,R.Screiner,D.Schmitz,M.Heuken CS Mantech 2012 growth of InGaAS Conference April 23th-26th Boston Massachusetts USA
[10] A.Ouerdane, M.Bouslama, A. Abdellaoui, M.Ghaffour Y. Al-Douri, Study of InGaAsN/GaAs material for optoelectronic applications First International Conference on Innovative Materials and Techniques CIMT November 12-15,2012 Hammamet (Tunisia)
[11] A.Nouri, Z.Lounis, A.Ouerdane, M.Ghaffour, M.Bouadi, H.Dumont, L. Auvray, M.Bouslama 2007 The behaviour of ternary compounds InGaAs and GaAsN subjected to electron irradiation, Vacuum **81** 979
[12] H. Baaziz , Z. Charifi , Ali Hussain Reshak , B. Hamad ,Y. Al-Douri 2012 Appl Phys A **106** 687
[13] Babar Bachir, 2009 Electronics/Telecommunications University of Gävle Supelec 592 35
[14] R.A. Abdullah, K. Ibrahim 2010 Optoelectronics and Advanced Materials–Rapid Com. **4** 568
[15] Jeffrey O. White et al. 2011 ARL-TN **1** 0451
[16] Safwat W.Z. Mahmoud 2007 Egypt Journal Solids **30** 277
[17] M. Aleshams 2009 Progress In Electromagnetic Research **8**121
[18] H.K. Liang, S. F. Yu and H.Y. Yang 2010 Applied Physics letters **97** 241107
[19] Vitezslav Jerabek, Ivan Huttel 2011 Radioengineering **20** 486
[21] Krishna Myneni 2011 Semiconductor Laser Rate Equation Solver **44** 1-37
[22] T. homas Erneux, Evgeny A.Viktorix and Paul Mandal 2007 Physical Review **A76** 0238
[23] Ovidio H. Anton, Dinesh Patel Carmen S. Menoni, Jeng-Ya Ych, T.T. Van Roy, L.J. Mawst, J.M.Pikal and Nelson Tansu 2005 Quantum Electronic **11** 1079
[24] Sheng Chu, Guoping Wang , Weihang Zhou, Yuqing Kong, Lin Li, Jingjian Ren and Jianlin Liu 2011 Nature Technology **6** 506
[25] H. A. McKay and R. M. Feenstra 2001, Journal Sci. **B19** 1644

Isomorphisim of generalized Holder spaces.

Abdumalik Rakhimov

International Islamic University Malaysia, Jalan Gombak, 53100 Kuala Lumpur

abdumalik@iium.edu.my

Abstract. In this paper we prove a theorem on isomorphism of the generalized Holder's spaces. A main result of the paper can be applied in the study of the spectral expansions in a closed domain.

1. Introduction and Preliminaries.

Among the partial differential operators elliptic type of operators are very important in many applications. One of the remarkable properties of elliptic operators is that the powers (including fractional powers) of their self-adjoin extensions are isomorphism in the spaces of differentiable functions (see for example in [1], [2]). Isomorphism properties of the elliptic operators and their powers serve not only for the solvability of the problems and uniqueness of the solutions but also provide integral representations of the classes of differentiable functions.

In this paper we consider self-adjoin extension of Laplace operator and isomorphism in functions space with generalized smoothness. A main result of the paper has an important application in proving sharpness of the eigenfunction expansions in the closed domains. Isomorphic properties are used to estimation of the regularized projectors associated with fractional powers of elliptic differential operators in different functional spaces (see in [3]).

Let $J(R^N)$ be the space of basic the Schwartch functions, $J'(R^N)$ be the space of linear continuous functionals on $J(R^N)$.

An operator

$$I_r f = F^{-1}\left[(1+|\xi|^2)^{\frac{-r}{2}} \cdot Ff\right]$$

is mapping $J'(R^N)$ to itself. Here F and F^{-1} are direct and inverse Fourier transformations in $J'(R^N)$, respectively. It is well known (see in [5]) that the operator I_r is isomorphism between the Besov spaces $B^r_{p,\theta}(R^N)$ $(1 \le p, \quad \theta \le \infty, \quad r > 0)$.

Let $a \ge 0$ be an integer. Assume that a function $\chi(t)$ is defined and continuous for $1 \le t \le \infty$, positive and vanishing in $t \to \infty$. Let $B^{a,\chi}_{p,\theta}(R^N)$ be generalized Besov's spaces.

It is well known that (see. in [4]) that an operator

$$I_{a,\chi} f = F^{-1}\left[(1+|\xi|^2)^{\frac{-a}{2}} \cdot \chi\left(\sqrt{1+|\xi|^2}\right) \cdot Ff\right]$$

is isomorphism in the Besov spaces $B^{a,\chi}_{p,\theta}(R^N)$, i.e.

$$B^{a,\chi}_{p,\theta} = I_{a,\chi} B^0_{p,\theta}$$

In this if $\chi(t) = t^{-\alpha}$ $(0 < \alpha \le 1)$, then obtain the statement mentioned above. In general case function χ must satisfy additional conditions.

Let Ω be N - dimensional bounded domain with sufficiently smooth boundary ($N \ge 2$) and let A be a self-adjoin extension of the operator $-\Delta$ in $L_2(\Omega)$. For any real number $\alpha > 0$ we define

$$(I + A)^{-\alpha} = \int_0^\infty (1 + \lambda)^{-\alpha} dE_\lambda ,$$

where $\{E_\lambda\}$ is a family of spectral projectors associated with the operator A. The operator $(I + A)^{-\alpha}$ is an integral operator with the kernels $K_\alpha(x,y)$. Functions $K_\alpha(x,y)$ are called fractional order kernels.

Definition 1. A positive continues function $\nu(t)$ in $t \ge 1$ is called weakly oscillating if for any $\varepsilon > 0$ there is $t_0 = t_0(\varepsilon)$ such that for all $t \ge t_0$ a function $t^{\varepsilon} \cdot \nu(t)$ is monotone increasing and the function $t^{-\varepsilon} \cdot \nu(t)$ monotone decreasing.

2. Statement of Main Results

Let $\chi(t)$ be a weakly oscillating function. Assume that the function $\chi(t)$ is enough smooth and functions $t^m\chi^{(m)}(t)$ also weakly oscillating.

The Bessel-McDonalds kernels are defined as

$$\Im_{\alpha,\chi}(r) = r^{1-\frac{N}{2}} \cdot \int_0^\infty t^{\frac{N}{2}} \cdot J_{\frac{N}{2}-1}(rt) \frac{\chi(\sqrt{t^2+1})}{\left(t^2+1\right)^{\frac{\alpha}{2}}} dt,$$

where J_ν is the Bessel function, $r = |x|$, $\alpha > 0$.

Let $\Phi(t) = \dfrac{\chi(\sqrt{t^2+1})}{(t^2+1)^{\alpha/2}}$. Then the operator $\Phi(A)$ is defined by following equality

$$\Phi(A) = \int_0^\infty \frac{\chi(\sqrt{\lambda^2+1})}{(\lambda^2+1)^{\alpha/2}} dE_\lambda, \quad \alpha > 0.$$

The operator $\Phi(A)$ is an integral operator with the kernel $T_{\alpha,\chi}(x,y)$ called a generalized kernel of fractional order.

For the fixed $h \in R^N$, we introduce a symmetric difference operator $\Delta_h f(x) = f\left(x+\frac{h}{2}\right) - f\left(x-\frac{h}{2}\right)$. and let $\Delta_h^n = \Delta_h \Delta_h^{n-1}$. For any function $u \in L_2(\Omega)$, a symbol $\omega_2(u,r)$ denotes second order module of continuity in the metric L_2, i.e.

$$\omega_2(u,r) = \sup_{|h|<r} \left\| \Delta_h^2 u \right\|_{L_2(\Omega_h)},$$

where $\Omega_h = \left\{ y \in \Omega : \ dist\left(y, \partial\Omega\right) < h \right\}.$

Let $\alpha > 0$, $\alpha = \nu + \zeta$, where ν is a non negative integer, $0 < \zeta \le 1$. By $H_\chi^\alpha(\Omega)$ we denote a space of functions from $L_2(\Omega)$ with the finite norm

$$\|\cdot\|_{\alpha,\chi} = \|\cdot\|_0 + \sum_{|\beta|=\nu} \sup_{0<\delta<1} \frac{\omega_2(D^\beta\cdot,\delta)}{\delta^\varsigma \cdot \chi\left(\frac{1}{\delta}\right)},$$

where $\| \ \|_0$ denotes norm in $L_2(\Omega)$.

Let $\Omega^0 \subset\subset \Omega$, then the symbol $\| \cdot \|_{\alpha,\chi,0}$ denotes a norm in $H_\chi^\alpha(\Omega^0)$. The main result of this paper is proving following

Theorem 1. Let $\Omega^0 \subset\subset \Omega$. Then there exists a constant c such that for any function $g \in L_2(\Omega)$ the estimate

$$\|\Phi(A)g\cdot\|_{\alpha,\chi,0} \leq c \cdot \|g\|_0. \tag{1}$$

is valid.

3. Proofs of Theorems 1.

Note that the function $T_{\alpha,\chi}(x,y)$ can be represented as

$$T_{\alpha,\chi}(x,y) = \Im_{\alpha,\chi}(x,y) + Q_{\alpha,\chi}(x,y),$$

where $Q_{\alpha,\chi}(x,y)$ is a continuous differentiable of the given order in Ω and for the Bessel-McDonald kernels the following estimations hold

$$\left| D^\beta \Im_{\alpha,\chi}(r) \right| \leq c \cdot \begin{cases} \cdot r^{\varsigma - N} \cdot \chi\left(\frac{1}{r}\right), & r_0 > r > 0 \\ r^{-\left(\frac{N-1}{2}+\varsigma\right)}, & r > r_0 \end{cases}. \tag{2}$$

Here $|\beta| = \nu$.

Accordingly the operator $\Phi(A)$ can be represented as

$$\Phi(A) = I_{\alpha,\theta} + Q_{\alpha,\theta}.$$

From the continuity of the functions $\frac{\partial^{\beta}}{\partial x^{\beta}} Q_{\alpha,\chi}(x,y)$ in the domain $\overline{\Omega^0} \subset \Omega$ for any multi-index β, we obtain that an integral operator $Q_{\alpha,\theta}$ is continuous from $H_{\chi}^{\alpha}(\Omega^0)$ to $L_2(\Omega)$.

Using methods developed in [5] and estimation (2) it is not difficult to obtain following statement

Lemma 1. The second order module of continuity of the Bessel-McDonald kernel in the metric L_1 satisfies the estimation

$$\omega_1\left(D^{\beta}\mathfrak{I}_{\alpha,\chi},\delta\right) \le c\cdot\delta^{\varsigma}\cdot\chi\left(\frac{1}{\delta}\right), \qquad (3)$$

where $|\beta| = \nu$.

Using the estimation (3) and the generalized the Minkowski inequality obtain the following estimation,

$$\omega_2\left(D^{\beta}I_{\alpha,\chi}g,\delta\right) \le c\cdot\delta^{\varsigma}\cdot\chi\left(\frac{1}{\delta}\right)\cdot\|g\|_0$$

for the second order module of continuity of function $D^{\beta}I_{\alpha,\chi}g(x)$ in metric L_2

Therefore,

$$\sum_{|\beta|=\nu}\sup_{0<\delta<1}\frac{\omega_2\left(D^{\beta}I_{\alpha,\chi}g\cdot,\delta\right)}{\delta^{\varsigma}\cdot\chi\left(\frac{1}{\delta}\right)} \le c\cdot\|g\|_0.$$

From this inequality it follows that integral operator $I_{\alpha,\chi}$ continuous from L_2 to L_2 . Theorem proved.

4. Acknowledgment.

This work done with the support of IIUM University Research Grant (Type B) EDW B12-397-0875

References

[1] Hans Triebel, Interpolation Theory - Function Spaces - Differential Operators, Wiley, Jan 6, 1999 - Science - 532 pages.

[2] S. A. Alimov, "Fractional powers of elliptic operators and isomorphism of classes of differentiable functions," Differentsial'nye Uravneniya, vol. 8, pp. 1609–1626, 1972 (Russian).

[3] V.A. Il'in, Spectral theory of differential operators, Moscow, (1990) 190 pages.

Energy Properties of rotational bands of $^{170-174}Yb$ isotopes

Abdurahim A Okhunov[1,3], **Anwer A M Al-Sammarrae**[2], **and G.Turaeva**[1]

[1] Department of Science in Engineering, International Islamic University Malaysia, 57528, P.O. 10, Kuala Lumpur, Malaysia

[2] Department of Physics, Faculty of Science, University of Malaya, 50603, Kuala Lumpur, Malaysia

[3] Institute of Nuclear Physics, Academy of Science of Uzbekistan, 100132, Tashkent, Uzbekistan

E-mail: abdurahimokhun@iium.edu.my , aaokhunov@gmail.com

Abstract.

Accounting for Coriolis mixing of experimentally known rotational bands non-adiabatic effects in energy of low-lying excited states are investigated, within phenomenological model. The Calculations for isotopes $^{170,172,174,176}Yb$, are carried out.

The energy and wave function structure of excited states are calculated. The finding reveals that the bands mixing has been found to have considerable impact on the wave function of low-lying states 0^+ and 2^+ bands.

1. Introduction

The medium and heavy mass isotopes of Ytterbium located in the rear-earth mass region are well deformed nuclei that can be populated to very high spin. Mach experimental information on even-odd mass of Yb isotopes has became more abundant [1]-[6].

A series of methods and nuclear models for describing the properties of the ground and excited positive - parity of even - even deformed nuclei are described and tested on practical problems. The adiabatic model of the nucleus proposed by Bohr and Mottelson, which is a phenomenological model, has played an important role in the study of the properties of deformed nuclei [7].

The nuclei $^{170,172,174}Yb$ have been well studied. It is important to note that these are investigated in a number of ways such as radioactive decay of $^{170,172,174}Lu$, and different nuclear reactions. In these isotopes, many 1^+ states and $K^\pi = 0^+, 2^+$ bands have been observed. For instance, the excited energy $K^\pi = 0^+_\ell(0^+_\ell)$ and $2^+_m(2^+_m)$ it rises with the increase in number of neutrons, except ^{176}Yb.

In the present paper we are investigated properties of the low - lying states of the isotopes Yb, within the phenomenological model [8, 9]. The calculation performed for the $^{170,172,174,176}Yb$ isotopes. The energy spectra and wave function of the excited states of ground band, $\beta_\ell-$, γ_m- bands are calculated.

2. The Model

To analyze the properties of low-lying positive parity states in Yb isotopes, the phenomenological model of [8] is exploited. This model takes into account the mixing of states of the $K^\pi = 0^+, 2^+$ and 1^+ bands. The Hamiltonian model is

$$H = H_{rot}(I^2) + H_{K,K'} \tag{1}$$

$$H^{\sigma}_{K'K}(I) = \omega_K \delta_{K,K'} - \omega_{rot}(I)(j_x)_{K,K'}\zeta(I,K)\delta_{K,K'\pm 1} \tag{2}$$

where ω_K– bandhead energy of rotational band, $\omega_{rot}(I)$– an angular frequency of rotational nucleus, $(j_x)_{K,K'}$– matrix elements which describe Coriolis mixture between rotational bands and

$$\zeta(I,0) = 1 \qquad \zeta(I,2) = \sqrt{1 - \frac{2}{I(I+1)}}.$$

The rotational part of Hamiltonian (1) $H_{rot}(I)$ is diagonal by wave functions. Note that $H_{rot}(I)$ is determined by exploiting Harris parameterization for energy and angular momentum [10]

$$E_{rot}(I) = \frac{1}{2}\Im_0\omega^2_{rot}(I) + \frac{3}{4}\Im_1\omega^4_{rot}(I) \tag{3}$$

$$[I(I+1)]^{1/2} = \Im_0\omega_{rot}(I) + \Im_1\omega^3_{rot}(I) \tag{4}$$

where $\Im_0$ and $\Im_1$– are the inertia parameters of the rotational core.

The rotational frequency of the core $\omega_{rot}(I)$ is found by solving cubic equation (4). This equation has two imaginary roots and one real root. The real root is as follows

$$\omega_{rot}(I) = \left\{\frac{\widetilde{I}}{2\Im_1} + \left[\left(\frac{\widetilde{I}}{2\Im_1}\right)^2 + \left(\frac{\Im_0}{3\Im_1}\right)^3\right]^{\frac{1}{2}}\right\}^{\frac{1}{3}} + \left\{\frac{\widetilde{I}}{2\Im_1} - \left[\left(\frac{\widetilde{I}}{2\Im_1}\right)^2 + \left(\frac{\Im_0}{3\Im_1}\right)^3\right]^{\frac{1}{2}}\right\}^{\frac{1}{3}} \tag{5}$$

where $\widetilde{I} = \sqrt{I(I+1)}$. Equation (5) gives $\omega_{rot}(I)$ at the given spin I of the core.

Solving the Shrödinger equation

$$H^{\sigma}_{K,K'}\Psi^I_{K,K'} = \mathcal{E}_K(I)\Psi^I_{K,K'}. \tag{6}$$

we define eigne function and energy of a Hamiltonian. The total energy of state is defined by

$$E_K(I) = E_{rot}(I) + \mathcal{E}_K(I) \tag{7}$$

3. Energy Spectra and Structure of States

The calculations have been carried out for the isotopes $^{170,172,174,176}Yb$. All experimentally known rotational bands of positive parity have been included in basis of Model Hamiltonian.

The experiment suggests that $m = 5$ band with $K^\pi = 0^+_m$, one band $\ell = 1$ with $K^\pi = 2^+_\ell$, and $\nu = 19$ with $K^\pi = 1^+_\nu$ states in ^{170}Yb [11]. These all $n = m + \ell + \nu = 25$ rotational bands have been included in the basis states of Hamiltonian (1). For the isotopes $^{172,174}Yb$, basis states of Hamiltonian include $n = 15$ ($m = 5$, $\ell = 2$ and $\nu = 8$) and $n = 22$ ($m = 5$, $\ell = 2$ and $\nu = 15$), correspondingly [12, 13, 14].

The parameters of inertia $\Im_0$ and $\Im_1$ are estimated by exploiting Harris parameterization (4), and using the experimental data for energy up to spin $I \leq 8\hbar$ for ground band [15].

The Hamiltonian (2) has transformational properties, that the state (3) can be classified as quantum number — $\sigma = \pm 1$ signature, which imposes restrictions on angular momentum values.

$$(-1)^I \sigma = 1.$$

For the states with negative signature $\sigma = -1$, Hamiltonian (2) has dimension $n = \ell + \nu$, as in bands with $K^\pi = 0^+_m$ the are no condition states with odd spins I. For the states with positive signature $\sigma = +1$, Hamiltonian (2) has dimension $n = m + \ell + \nu$.

The model parameters are described as follows:

a) the bandhead energy ground ω_{0_1} and $K^\pi = 0^+_m$ bands has taken from experiment, as they are not revolted by Coriolis force. Bandhead energy of 1^+_ν bands are also defined from an experiment [11, 12]

$$\omega_{1_\nu} = E^{exp.}_{1_\nu}(I = 1) - E_{rot}(I = 1);$$

b) matrix elements $(j_x)_{2_\ell 1} = (j_x)_{2_\ell 1_\nu}$ and bandhead energy of 2^+_ℓ– bands ω_{2_ℓ} are determined from the most favored experimental and theoretical spectrum of energy states with a negative signature $\sigma = -1$, e.a. for energy state for even spins I;
c) the matrix elements $(j_x)_{0_m 1} = (j_x)_{0_m 1_\nu}$ defined by the least square method from the best fitted of theoretical energy spectra state with positive signature $\sigma = +1$ with experimental data.

The obtained values of model parameters are presented in Table 1.

Table 1. The used in the calculations for the isotopes $^{170-176}Yb$.

A	$\jmath_0$	$\jmath_1$	ω_{2_1}	ω_{2_2}	$(j_x)_{gr,1^+}$	$(j_x)_{\beta_1,1^+}$	$(j_x)_{\gamma_1,1^+}$
170	35.064	87.814	1.075	–	0.186	0.393	0.727
172	37.422	90.176	1.389	1.540	0.275	0.977	0.325
174	38.556	84.951	1.556	2.652	0.185	0.400	0.085
176	35.945	66.448	1.2109	–	0.200	0.540	0.090

Whre: $\jmath_0, \jmath_1$ are inertial parameters of rotational core $\left(\frac{\hbar^2}{MeV}, \frac{\hbar^4}{MeV^3}\right)$, ω_K– are parameters of bandhead energy (MeV), and $(j_x)_{K',K}$– are matrix elements of Coriolis mixture.

In the figures 1-4 the comparison of calculated values of energy of low-lying states for isotopes of $^{170-176}Yb$ with experimental data [11, 13, 14, 17] and [18] and with taken results by $IBM-1$ model are presented. From the comparison we can, that the model qualitatively reproduces experimental energy of rotational states up to energy $3MeV$. However, in high spin values $I \geq 10\hbar$ noticeable deviation has been observed in calculated values of energy and that obtained from experiment. Note that this deviation increases with the growth of angular momentum I. This is probably due to the fact that the influence of rotation on internal nuclei structure has not been considered in this model. In future, we will study electromagnetic properties of low-lying states $I < 10\hbar$.

Furthermore the model predicts the energies, spin and parity of β_2–, β_3–, β_4–, γ_2– and 1^+ band states and as it shown in the Tables 2 – 6 for isotopes $^{170-176}Yb$, respectively. In these tables our calculations are in good agreement with the experimental data.

The structure of the wave functions of states $\Psi_{K',K}$ for $K^\pi = 2^+_1$– and 0^+_2– bands for $^{170,172,174,176}Yb$, are provided in Fig. 5–8, respectively. The components which have small values are not illustrated in Figures.

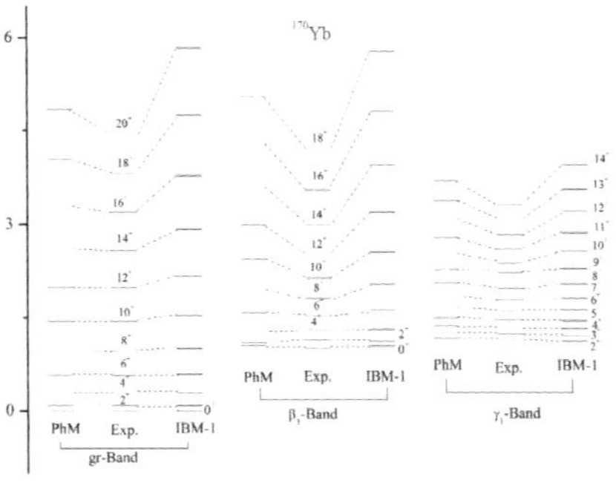

Figure 1. The comparison of calculation energy values by PhM and $IBM-1$ with experimental data for ^{170}Yb.

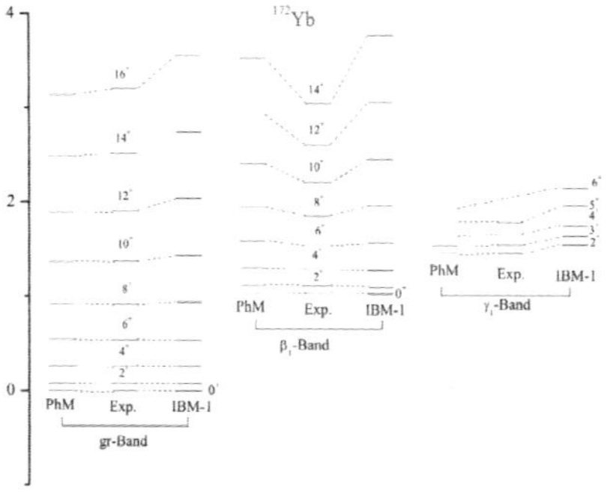

Figure 2. The comparison of calculation energy values by PhM and $IBM-1$ with experimental data for ^{172}Yb.

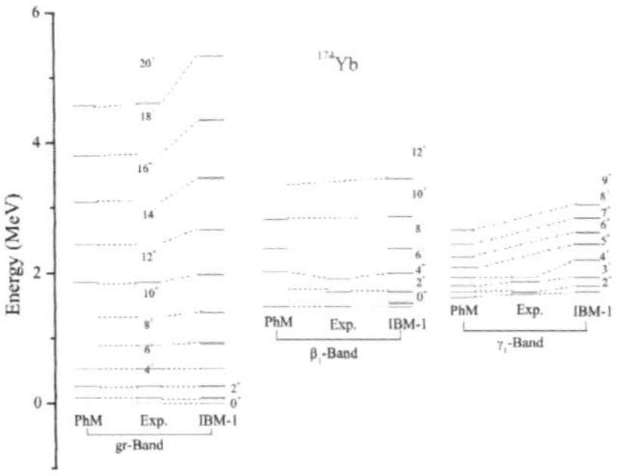

Figure 3. The comparison of calculation energy values by PhM and $IBM-1$ with experimental data for ^{174}Yb.

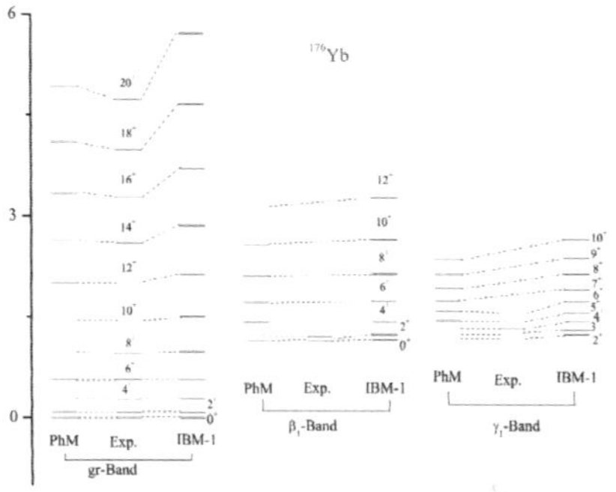

Figure 4. The comparison of calculation energy values by PhM and $IBM-1$ with experimental data for ^{176}Yb.

Table 2. The energy levels of β_2- band of Yb isotopes (in MeV).

	^{170}Yb		^{172}Yb		^{174}Yb		^{176}Yb	
I	Exp.[17, 11]	PhM	Exp.[17, 13]	PhM	Exp.[17, 14]	PhM	Exp.[17, 18]	PhM
0^+	1.228	1.228	1.404	1.404	1.885	1.885	1.518	1.518
2^+	1.306	1.313	1.476	1.483	1.958	1.962	1.610	1.601
4^+	–	1.507	1.632	1.666	2.123	2.140	–	1.792
6^+	–	1.804	–	1.947	–	2.414	–	2.086
8^+	–	2.195	–	2.317	–	2.770	–	2.476
10^+	–	2.669	–	2.769	–	3.221	–	2.954
12^+	–	3.220	–	3.295	–	3.740	–	3.512

Journal of Physics: Conference Series **553** (2014) 012011 doi:10.1088/1742-6596/553/1/012011

Table 3. The energy levels of β_3- band of Yb isotopes (in MeV).

	^{170}Yb		^{172}Yb		^{174}Yb		^{176}Yb	
I	Exp.[17, 11]	PhM	Exp.[17, 13]	PhM	Exp.[17, 14]	PhM	Exp.[17, 18]	PhM
0^+	1.479	1.479	1.794	1.794	2.113	2.110	1.779	1.779
2^+	1.534	1.564	1.849	1.873	2.172	2.178	–	1.862
4^+	1.667	1.758	1.975	2.056	2.336	2.356	–	2.053
6^+	–	2.055	2.156	2.156	–	2.630	–	2.347
8^+	–	2.446	–	2.707	–	2.993	–	2.737
10^+	–	2.920	–	3.159	–	3.437	–	3.215
12^+	–	3.471	–	3.685	–	3.956	–	3.773

Table 4. The energy levels of β_4- band of Yb isotopes (in MeV).

	^{172}Yb		^{174}Yb	
I	Exp.[17, 13]	PhM	Exp.[17, 14]	PhM
0^+	1.894	1.894	2.821	2.821
2^+	1.956	1.973	–	2.898
4^+	2.100	2.156	–	3.076
6^+	–	2.437	–	3.350
8^+	–	2.807	–	3.713
10^+	–	3.259	–	4.157
12^+	–	3.785	–	4.676

Table 5. The energy levels of γ_2- band of Yb isotopes (in MeV).

	^{172}Yb		^{174}Yb	
I	Exp.[17, 13]	PhM	Exp.[17, 14]	PhM
2^+	1.608	1.619	2.728	2.727
3^+	1.700	1.698	2.793	2.804
4^+	1.803	1.802	2.882	2.905
5^+	1.926	1.931	–	3.031
6^+	2.075	2.083	–	3.179
7^+	–	2.257	–	3.350
8^+	–	2.453	–	3.542
9^+	–	2.669	–	3.754
10^+	–	2.905	–	3.986
11^+	–	3.159	–	4.236
12^+	–	3.431	–	4.505
13^+	–	3.720	–	4.791
14^+	–	4.026	–	5.093

Table 6. The energy levels of 1^+- band of Yb isotopes (in MeV).

	^{170}Yb		^{172}Yb		^{174}Yb		^{176}Yb	
I	Exp.[17, 11]	PhM	Exp.[17, 13]	PhM	Exp.[17, 14]	PhM	Exp.[17, 18]	PhM
1^+	1.606	1.605	2.009	2.006	1.624	1.605	1.819	1.818
2^+	1.832	1.662	2.047	2.059	1.674	1.657	1.867	1.874
3^+	–	1.746	2.109	2.138	1.733	1.734	–	1.956
4^+	–	1.856	2.193	2.242	1.859	1.835	–	2.065
5^+	–	1.993	–	2.371	–	1.961	–	2.200
6^+	–	2.153	–	2.523	–	2.109	–	2.359
7^+	–	2.337	–	2.697	–	2.280	–	2.542
8^+	–	2.544	–	2.893	–	2.472	–	2.749
9^+	–	2.771	–	3.109	–	2.684	–	2.977
10^+	–	3.018	–	3.345	–	2.916	–	3.227
11^+	–	3.285	–	3.599	–	3.166	–	3.496
12^+	–	3.569	–	3.871	–	3.435	–	3.785
13^+	–	3.871	–	4.160	–	3.721	–	4.093
14^+	–	4.190	–	4.466	–	4.023	–	4.419

From Fig. 5–8, we can see that $K^\pi = 0_2^+$ and $K^\pi = 2_1^+$ bands states in ^{170}Yb, $K^\pi = 0_3^+$ and $K^\pi = 2_1^+$ bands in ^{172}Yb are mixed strongly even in low spin values I. In case ^{174}Yb and ^{176}Yb, $K^\pi = 0_2^+$ and $K^\pi = 2_1^+$ bands states are very close, but the mixture effect is not so strong. In isotopes $^{170,172}Yb$, considerable deviation in signature of the states $K^\pi = 2_1^+$ band can be observed. This reflects in the values of probability electromagnetic transitions. In this case, description of quantum number K is difficult for these states. Thus, in ^{170}Yb, a number of research works [17] in this context note that the states with $I = 2^+$ (1.1386 MeV) and $I = 2^+$ (1.1454 MeV) $K = 2$ and $K = 0$, respectively. On the other hand, some works [19] document that $K = 0$ and $K = 2$, correspondingly.

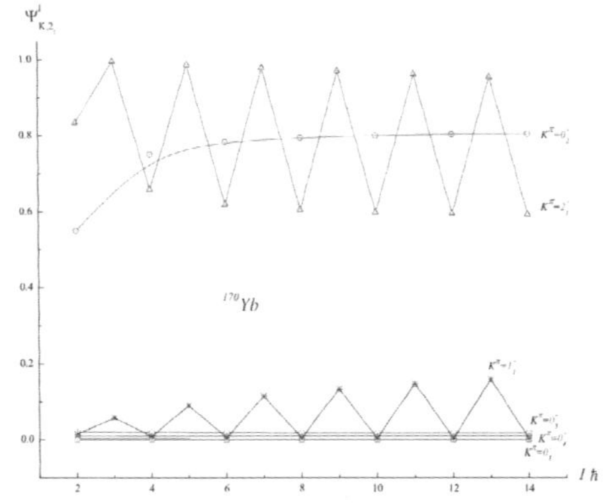

Figure 5. The structure of the wave function of γ_1 band for ^{170}Yb.

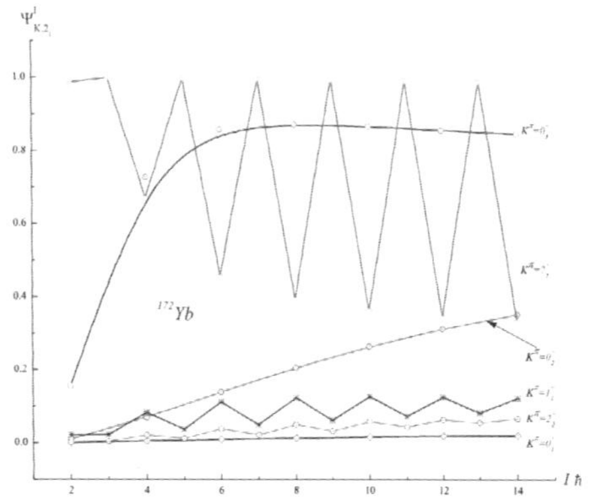

Figure 6. The structure of the wave function of γ_1 band for ^{172}Yb.

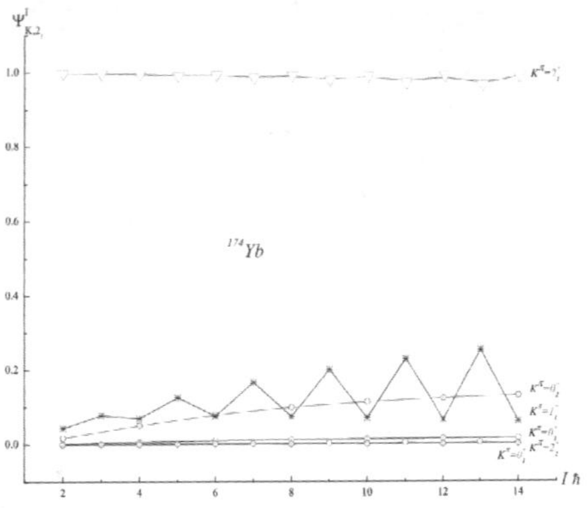

Figure 7. The structure of the wave function of γ_1 band for ^{174}Yb.

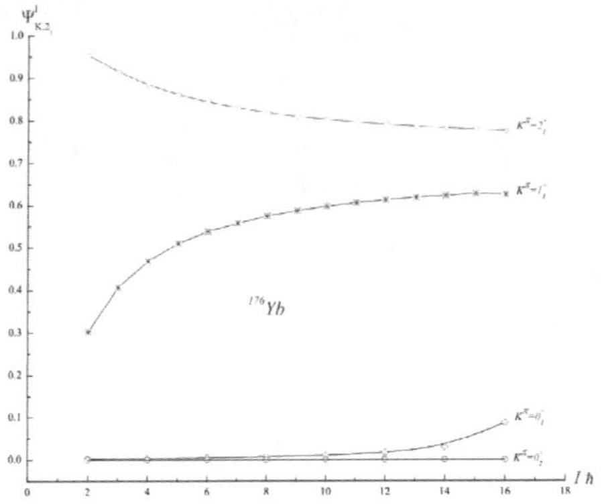

Figure 8. The structure of the wave function of γ_1 band for ^{176}Yb.

4. Conclusion

In the present work, non-adiabatic effects in energies and wave function structure of low-lying excited states are studied within the phenomenological model which taking into account Coriolis mixing of all experimentally known rotational bands for isotopes $^{170,172,174,176}Yb$.

The energy and structure of wave functions of low-lying excited states of $^{170-176}Yb$ isotopes are calculated. The finding reveals that the bands mixing has been found to have considerable impact on the wave function of low-lying states 0^+ and 2^+ bands.

The predicted low-lying levels ($gr-, \beta_1-$ and γ_1- band) by PhM are in good agreement with the experimental data as compared with those by $IBM-1$ for all nuclei of interest. In addition, the PhM is successful in predicting the β_2-, β_3-, β_4-, γ_2- and 1^+- band while $IBM-1$ fails.

Acknowledgments

This work has been financial supported by IIUM University Research Grant (Type B) EDW B13-034-0919 and MOHE Fundamental Research Grant Scheme FRGS13-074-0315. We thank the Islamic Development Bank (IDB) for supporting this work under scholarships 37/IRQ/P30. The author A. A. Okhunov is grateful to Prof. Ph.N. Usmanov for useful discussion and exchange ideas.

References

[1] Mo J N and et al. 1987 *Nucl. Phys. A* **472** 295
[2] Jonsson S, Roy N, Ryde H, Walus W and et al. 1986 *Nucl. Phys. A* **449** 537
[3] Beck E M, Bcelar J C, Deleplanque M A, Diamond R M and et al. 1987 *Nucl. Phys. A* **464** 472
[4] Byrski T and et al. 1987 *Nucl. Phys. A* **474** 193
[5] Fewell M P and et al. 1988 *Phys. Rev. C* **37** 101
[6] Granja C, Posíšil S, Chrien R E and Telezhnikovc S A 2005 *Nucl. Phys. A* **757** 287
[7] Bohr A and Mottelson B R 1997 *Nuclear Structure Vol.* **1.2** Benjamin New York
[8] Usmanov Ph N, Okhunov A A, Salikhbaev U S and Vdovin A 2010 *Phys. Part. Nucl. Lett.* **7** 185
[9] Usmanov Ph N, and Mikhailov I N 1997 *Phys. Part. Nucl. Lett.* **28** 348
[10] Harris S M 1965 *Phys. Rev. B* **138** 509
[11] Baglin C M 2002 *Nucl. Data Sheets* **96** 611
[12] Zilges A, von Brentano P, Wesselborg C and et al. 1990 *Nuc.Phys. A* **507** 399
[13] Singh B 1995 *Nucl. Data Sheets* **75** 199

[14] Browne E and Junde H 1999 *Nucl. Data Sheets* **87** 15
[15] Okhunov A A, Hasan Abu Kassim and Usmanov Ph N 2011 *Sains Malaysiana* **40** No.1 1
[16] Reich C W, Greenwood R C, Lokken R A 1974 *Nucle. Phys. A* **228** 365
[17] Begzhanov R B,Belinkiy V M, Zalyubovskiy I I and Kuznichenko A B 1989 *Handbook on Nuclear Physics* Vol.**1,2** Tashkent FAN
[18] Basunia M S 2006 *Nucl. Data Sheets* **107** 791
[19] Baglin C M 1996 *Nucl. Data Sheets* **77** No.2 125

AUTHOR INDEX

Institute of Physics
Dirac House, Temple Back
Bristol BS1 6BE UK

CURRAN ASSOCIATES INC
proceedings.com